THE IMPACT OF RISING OIL PRICES ON THE WORLD ECONOMY

One is tempted to say that after the first great oil price increase in 1973 the world economy witnessed a further round of eating the apple of the tree of knowledge and that we are still digesting. The reorientation of economic research activities, prompted by the rising trend of real energy prices, has manifested itself in many if not most branches of economic analysis. The papers in this volume can also be seen as a reflection of this reorientation.

Those in the first half of the volume are mainly devoted to empirical assessment of the economic impact of rising oil prices. Large econometric models developed within the United Nations, OECD, the International Energy Agency and in the United States are used as the tools of analysis. Among the topics covered in this part are the regional effects of energy saving, the relationship between energy prices and productivity growth, the consequences of energy price decontrol, the macroeconomic impact of rising oil prices in small open countries and the importance of the OPEC respending rate.

In the second half of the volume the emphasis is on theory and in particular on the economics of exhaustible resources. A central theme is the optimal production and price policies for a resource such as oil under various market structures, but a wide variety of additional aspects are covered in the analysis. Among these are the incentives for developing resource substitutes (backstop technologies), uncertainty about the date of discovery of a substitute, the timing of innovations and the concept of scarcity in extractive resource markets. Other contributions deal with the trade-off between efficiency in production and *ex post* flexibility in input proportions and the effects of rising oil prices in a two-sector economy with immobile labour.

Lars Matthiessen is Professor of Economics at Odense University, Denmark. He has a mixed Danish and Swedish background. After graduation from Copenhagen University in 1957 he settled in Sweden where he worked at the National Institute of Economic Research and as a Lecturer at the Stockholm School of Economics. He was a Visiting Professor at Columbia University, New York (1969–70). In 1973 he defended his dissertation *A Study in Fiscal Theory and Policy* in Stockholm. The following year he became a member of the Swedish Business Cycle Council, which started its activities that year. In 1975 Lars Matthiessen returned to Denmark and took up his present appointment. He has been a member of the chairmanship of the Danish Economic Council (1976–81) and editor of *The Scandinavian Journal of Economics* (1980).

THE IMPACT OF RISING OIL PRICES ON THE WORLD ECONOMY

Edited by

Lars Matthiessen
Odense University, Denmark

This collection was originally published in
The Scandinavian Journal of Economics, Vol. 83, 1981, No. 2

First published in book form 1982 by
THE MACMILLAN PRESS LTD
London and Basingstoke
Companies and representatives
throughout the world

ISBN 0 333 33185 0

Printed in Hong Kong

CONTENTS

INTRODUCTION

Important peacetime changes in society are normally too slow and gradual to be discernible while they occur. When seen in this perspective, developments in the oil markets since 1973 have been less than normal. One is tempted to say that in 1973 the world economy witnessed a further round of eating the apple from the tree of knowledge and that we are still digesting. Although the economics (and neighboring) professions have had eight years for reflection and research, the process of identifying and assessing the impact of rising energy prices on the world economy is likely to go on for some time to come. Needless to say, such an analysis is complicated by the fact that the rising trend of energy prices has brought about changes in important mechanisms in the world economy (including the process of economic policy-making).

The reorientation of economic research activities, prompted by the rising trend of real energy prices, has manifested itself in many if not most branches of economic analysis. More formal signs of the change in emphasis are e.g. the addition, in June 1977, of energy as a specific group (number 723) to the classification system of the *Journal of Economic Literature* and the appearance of specialized journals on energy economics. The papers in this special issue of *The Scandinavian Journal of Economics* can also be seen as a reflection of this reorientation.

Part I contains six papers which mainly, but not at all exclusively, are devoted to empirical assessment of the impact, in various respects, of rising oil prices. In all cases large econometric models are used for simulations. Part II comprises six papers which are mainly theoretical. Four of them reflect the renewed awareness of the problems associated with the exhaustibility of natural resources. The next few pages contain brief introductions to the twelve papers.

In her article *Anne P. Carter* explores some of the consequences for the world economy of an energy conservation program entailing reductions in energy consumption of 30–50 percent over the next 20 years in different regions of the world. The results reported are derived from simulations of the UNITED NATIONS WORLD MODEL, a very large input–output model of the world economy, which was constructed with the assistance of the author. For the oil-exporting countries in the Middle East, the assumed conservation program would result in a 50 percent reduction in their net export of petroleum. This of course means that these countries' oil reserves are depleted less rapidly. For the developed market economies it is shown that conservation in general leads to increased investment at the expense of consumption with only minor effects on overall GDP. The major benefit of conservation to developed regions is a significant reduction in the balance-of-payments deficit. Since prices are exogenous in the world model, the effect of conservation on future oil prices cannot be estimated in this model.

The following two articles are concerned with the American economy. In his paper *Dale W. Jorgensen* discusses the relationship between energy prices and productivity growth in the United States during the period 1948–76. The analysis is based on an econometric model of sectoral growth and data for 35 individual industries. The inputs considered are capital, labor, energy and materials input. It is shown that the drastic decline in the average rate of output growth from 3.5 percent in 1948–76 to 0.9 percent in 1973–76 is due mainly to a fall in productivity growth of 1.8 percentage points (from 1.1 percent to −0.7 percent). It is further concluded that the observed fall in productivity growth cannot be explained entirely by reallocation of output and inputs among sectors. Instead it is concluded that this fall is partly a consequence of the sharp increase in the price of energy relative to other productive inputs, which occurred after 1973 despite the use of price controls. Considering the performance of the U.S. economy since 1973, it is stressed that we can anticipate a further slowdown of productivity and output growth for the American economy as a whole. In order to avoid such a development, measures which reduce the price of capital and labor inputs are seen as essential.

A controversial issue in the United States in recent years has been whether energy price control should be continued or abolished. The economic effects of energy price decontrol are analyzed quantitatively in *Edward A. Hudson's* contribution. The analytical framework used is the Hudson–Jorgensen ENERGY–ECONOMY MODEL, which is a structural growth model of the U.S. economy. Three major effects of the energy price increases associated with decontrol are recognized: productivity effects of changes in energy demand and in energy supply and the effect of induced international trade adjustments. Demand side substitutions will result in a less energy-intensive product mix produced by less energy-intensive methods. This implies productivity decreases. These will outweigh the productivity increases brought about by a greater supply of low cost domestic energy. The international trade effects are associated with the appreciation of the dollar due to reduced energy imports. This will in turn lead to a reduction in exports. This development permits increased domestic demand and productivity increases far greater than the net productivity loss mentioned above. Thus decontrol can be assumed to have a favorable effect on total production in the United States.

The Nordic countries have been affected in different ways by developments in oil markets. Norway is the only large-scale oil producer and in addition, Norwegian hydro-electric power production covers more than half of total energy requirements. Despite Sweden's large production of nuclear and hydro-electric power, and despite Denmark's and Finland's extensive and increasing use of coal, these three countries rely heavily on imported oil from the Middle East and—in the case of Finland—from the USSR.

One would thus expect the macroeconomic impact of the 1979/80 oil price increase to be rather different in the four Nordic countries. The extent to which this is actually the case is shown in *Ian Lienert's* paper. His analysis is based on the OECD INTERLINK model of the world economy, which means that the effects of trade adjustments involving third countries are taken into account. It is assumed that economic policies are unchanged. A striking result of this analysis is that the effect on Finland's GDP is zero; the main reason is that Finland's terms-of-trade loss is largely offset by increased exports to the USSR. For the other three countries, the estimated GDP loss by 1981 is 3–4 percent. The fact that this also applies to Norway is due primarily to the assumption that additional public oil revenues are not respent. Not surprisingly, the balance-of-payments effects were strongly negative for Denmark and Sweden, slightly negative for Finland and strongly positive for Norway.

The quantitative results obtained in Ian Lienert's paper are of course sensitive to the assumption regarding the rate of OPEC respending. This topic is discussed in the following article by *Jan F. R. Fabritius* and *Christian Ettrup Petersen*. Their analytical framework is CRISIS, a world trade model system which is very similar to the OECD INTERLINK model. An operational measure of OPEC respending is developed and used to compare the rate of respending after the 1973/74 and 1979/80 oil price increases. It appears, perhaps contrary to certain expectations, that respending has been significantly lower after the second wave of oil price increases than after the first. The simulation results indicate that the rate of respending is of crucial importance for the impact oil price increases on economic activity in the industrial market economies. It also appears that Europe is more sensitive to variations in OPEC respending than the United States. It is concluded that although the two oil price shocks were of the same magnitude (the terms-of-trade loss was about 2 percent of GDP in either case), a deeper recession in Europe should be expected this time than after 1973/74 because of lower respending.

In the last of the empirically oriented papers, written by *Per-Anders Bergendahl* and *Clas Bergström*, the analytical context is not a complete model of the economy but the MARKAL model of a nation's energy system. This model, developed for use at the International Energy Agency, is a multiperiod, multi-objective linear programming model. It includes a physical representation of technologies, energy, flows and conversion efficiencies at various stages from extraction or import to the ultimate goal of satisfying the demand for useful energy. Energy demand is treated as exogenous and the model is thus unable to catch the interplay between the energy system and the rest of the economy. By the end of 1979, analyses using the MARKAL model had been completed in 15 countries. The MARKAL model is of course useful for a great number of purposes. One such purpose could be to identify trade-offs between various objectives in order to obtain a basis for policy evaluation. The trade-off between the total cost of energy and oil imports for Sweden for the period 1980–2020 is treated in the paper. It is also indicated how interfuel substitution, with alternative priorities, changes the optimal mix of the various energy raw materials in Sweden between 1990 and 2010.

The paper by *Karl-Göran Mäler* and *Lars Bergman* deals with a major problem in Sweden's (and other countries') energy policy, i.e. the economic risk associated with a high degree of dependence on imported oil combined with uncertainty about future oil prices. The authors consider a possible trade-off between static efficiency and flexibility *ex post* (that is, after the plant is built). The available technologies may be such that *ex post* flexibility of oil input coefficients can be attained only at the expense of static oil efficiency—and vice versa. If oil prices rise to unexpectedly high levels, large *ex post* flexibility would permit a reduction in oil inputs and thus lower the real income loss due to higher oil prices. This benefit should then be seen in relation to the cost, in the form of decreased static efficiency, associated with a flexible technology. Evidently the optimal mix of efficiency and flexibility will depend on the probability distribution of future oil price developments and the cost of additional flexibility. Some numerical experiments are carried out within the context of an MSG-model of the Swedish economy in order to obtain a rough idea of the social benefits of a flexible design of capital equipment. It is concluded that investments in increased flexibility may yield substantial benefits if uncertainty about future oil prices is sufficiently great.

In most short-run macroeconomic two-sector models, perfect labor mobility between the sectors is assumed. In his paper, *Michael Hoel* analyzes the effect of increased oil prices in an economy consisting of a tradeables and a non-tradeables sector, where there is no labor mobility between the sectors and where wages are determined in accordance with a "wage leadership" assumption. He also discusses the implications of rising oil prices for economic policies aimed at full employment. The question as to whether taxes should be raised or lowered is seen to depend on which of the two sectors has unemployment.

The last four papers are all contributions to the theory of exhaustible resources and complement each other very well. *Robert S. Pindyck* studies the optimal behavior of a producer of a resource such as oil. It is assumed that the price follows a growth path exogenously determined by a cartel (OPEC) and is subject to stochastic fluctuations around the growth path. It is also taken into account that marginal extraction costs are usually not constant, but vary with the rate of production and (inversely) with the level of reserves. Uncertainty regarding the future price is shown to affect the current rate of production in several ways. With a nonlinear marginal extraction cost function, stochastic price fluctuations will lead to increases or decreases in cost over time. This means that costs can be reduced by speeding up or slowing down the rate of production. The fact that a resource reserve has an "option" value may also provide an incentive for holding back production, which is speeded up as the uncertainty about future prices becomes greater.

In his explorative paper, *Partha Dasgupta* analyzes a series of problems related to the pricing and depletion of natural resources under oligopolistic market conditions. One of the issues discussed is the uncertainty of property rights due to a threat of expropriation by host countries. It is concluded that foreign extractors add a risk premium to the required rate of return and that the result is excessive depletion and too low a resource price as compared to the efficient outcome. A central theme in the paper is the relationship between the pricing of exhaustible resources and the incentives for developing technologies which would release an economy from binding resource constraints. It is emphasized that if a resource cartel (such as OPEC) sets a high price for its resource, this will dampen the incentives for R & D expenditures by rival firms. On the other hand, a price which is too high implies that revenues are initially too low. A delicate balance of these two considerations will determine the cartel's price policy. It is also concluded that a resource cartel has greater incentives than anyone else for acquiring the patent of a rival technology. Backstop technology may of course be developed within public enterprises as well as by private firms. Concerning public R & D expenditures, it is concluded that a marginal social cost-benefit analysis is insufficient in order to determine the optimal R & D outlay. In general, global social cost-benefit analysis is required.

Joseph E. Stiglitz and *Partha Dasgupta* analyze the rate of extraction of a natural resource prior to the discovery of a substitute, when the date of discovery is uncertain. In the analysis the markets for the resource and the substitute are alternatively assumed to be competitive and monopolistic. This gives five alternative market structures and the resulting prices for a given stock of the resource are compared. Evidently the rate of extraction of the resource will be lower, the higher the price of the resource. It is also clear that if the pre-invention resource price is high, the actual production of the substitute will be delayed. It is now shown that the lowest resource price will be charged in the case where competition prevails in the markets for both the resource and the substitute. This is hardly surprising. It is also shown that the highest price is obtained in the duopoly case where the resource is controlled by one monopolist and the substitute by another. The intermediate case is shown to be that of a single monopolist who controls both the resource and the substitute. The resource price under a pure monopoly of this type will be lower (higher) than in the case with a monopoly resource and a competitive substitute (a monopoly substitute and a competitive resource).

An important issue is resource scarcity and *Geoffrey Heal* devotes a large part of his article to this topic. It is emphasized that the development of resource prices is hardly an appropriate measure of the degree of scarcity. A related question is whether resources are used efficiently and it is discussed whether this is the case under monopoly and oligopoly. Assuming market equilibrium and accurate expectations, it is concluded that on purely theoretical grounds it does not seem likely that imperfect competition is a major cause of losses due to inefficiencies. Another issue examined in this paper is whether inefficiencies are to be expected in markets where information is less than full and where the equilibrating forces are too weak to secure the attainment of equilibrium. Not surprisingly, there are no clear-cut conclusions reached concerning the behavior of markets out of equilibrium because a very wide range of outcomes is possible.

It is hoped that the reader will find the treatment in the following papers as stimulating as the Editor did.

The generous financial support from the Swedish Energy Research and Development Commission for this special issue of *The Scandinavian Journal of Economics* is acknowledged with gratitude.

Lars Matthiessen

INTERNATIONAL EFFECTS OF ENERGY CONSERVATION*

Anne P. Carter

Brandeis University, Waltham, Massachusetts, USA

Abstract

This paper reports on an energy conservation scenario computed by the United Nations World Model. Energy prices are assumed to rise as they did in the base runs but shifts in fuel mix and energy-saving are assumed to be achieved through increased labor and capital costs. Results suggest that vigorous energy conservation would substantially decrease, but not eliminate, the payments stresses in the reference case. Regional effects vary greatly. Payments-constrained regions could develop somewhat faster despite additional requirements for investment goods. Developed economies would achieve improved trade balances through higher investment and some consumption sacrifice.

I. Introduction

Increases in energy prices since 1972 have jolted the world economy and there is general agreement that energy consumption will have to be curbed to restore balance. Energy conservation, decreased energy intensity of consumption and production, will, in turn, have major economic ramifications with different effects in each region of the world. This article explores some consequences of energy conservation for various sectors and regions of the world economy over the next 20 years in the context of the United Nations World Model.

Adjustments to changing energy supply conditions can be expected to include price and income responses. Rising energy prices are expected to lead to substitutions and induced technological changes that reduce the energy intensiveness of production and consumption. Regional incomes are affected by rising real costs of imported and domestic energy and changes in income also influence energy use. Ideally, one might hope to study the adjustment process in terms of a model that subsumes both price and income responses. Actually, the United Nations model does not allow for endogenous price response; changes in input and consumption structure must be introduced exogenously. Income and other effects of such exogenous changes can then be traced by solving the model.

---

* Alan Siu made most of the computations. The author is responsible for any errors.

The present study begins with a "scenario" for energy-saving structural change based on an independent analysis published by the Cavendish Laboratories for the World Energy Conference (1978). The Cavendish conservation scenario is compared with the reference scenario, AA, of the United Nations World Model and the latter's parameters are modified to form a World Model conservation scenario, AB. Solutions of the World Model before and after the changes are compared to analyze the effects of conservation on the regional and sectoral variables of the system.

Exogenous treatment of substitution and structural change has advantages for addressing long range energy questions. There is very little historical basis for econometric estimation of the long run responsiveness of the world economy to changes in energy prices. Prior to 1973, energy prices followed gradually declining trends, providing no adequate evidence for gauging responses to an abrupt price increase. Furthermore, energy use depends on the design of equipment in place and long lags in response to rising energy prices are the norm. Thus, the brief experience of price response in the seventies provides little guidance as to what may develop in the future.

The present "scenario" approach taps the expertise of specialists most likely to be in a position to anticipate future supply conditions and to appraise prospects for technological adaptation. Thus far, even specialists have only very general ideas about the form that conservation will take. As the concrete details of conservation strategies become clearer, additional simulations can incorporate them for further analysis.

As in most empirical work, conclusions are unavoidably sensitive to the structural features of the model and to the exogenous assumptions introduced. The most important of these are outlined in Sections I.1 and I.2.

I.1. The United Nations World Model

The United Nations World Model is described in several other sources, for example Carter & Petri (1977) and Leontief, Carter & Petri (1977), and space limitations permit only a brief outline of the model's structure here. It consists of four systems, one for the base year, 1970, and one each for 1980, 1990 and 2000. Each of the last three systems is linked to the previous decade's system through "history variables" that specify the amounts of capital goods (machinery and buildings) and nonrenewable mineral reserves available at the end of the previous decade. Capital requirements are determined by means of capital coefficients specified for each sector. Annual rates of investment are the sum of replacement requirements and a proportion of the difference between capital required in the given decade and the stock "inherited" from the previous decade. Both investment and resource decumulation is assumed to take place continuously throughout each decade.

The model divides the world into 15 regions and each regional submodel

Table 1. *Assignment of constraints to regions in the control scenario* (*AA*)

Constraint	Region	Basis for exogenous values assigned
Labor	North America	Projected labor force
	Western Europe (high income)	Projected labor force
	Soviet Union	Projected labor force
	Eastern Europe	Projected labor force
	Japan	Projected labor force
	Oceania	Projected labor force
Foreign exchange	Latin America (medium income)	Export earnings × 5
	Western Europe (medium income)	Export earnings × 5
	Asia (low income)	Export earnings × 2
	Arid Africa	Export earnings × 5
	Southern Africa	Export earnings × 2
Savings	Latin America (low income)	1970 excess savings (%)
	Asia, centrally planned	1970 excess savings (%)
	Tropical Africa	1970 excess savings (%)
Absorptive capacity	Middle East	9.5 % GDP growth rate

consists of 175 linear equations in 225 unknowns. Each region is represented by input–output, consumption and trade equations, with coefficients appropriate to income level, resource endowment and specialization history. Macroeconomic variables such as gross domestic product (GDP), consumption, investment and balance of payments are also computed for each region. Most of the variables of the system are measured in constant (1970) dollars but minerals and some other variables are measured in physical units. The "balance of payments" variable is computed as the sum of exports minus imports valued in *current* dollars and long term capital and aid flows projected in terms of pre-OPEC trends. The balance of payments provides an estimate of the regional foreign exchange surplus or shortfall to be covered by short term lending or by additional long term capital flows. Since this "balance" is measured in current dollars it is sensitive to projected prices of traded goods and particularly to changing energy prices.

The regional submodels are "rectangular" and some variables must be set exogenously. The choice of exogenous variables is varied from region to region to represent different modes of determining regional income level. Table 1 names the fifteen regions and indicates, for each, the constraint that determines income in the reference scenario, AA. The choice of constraints is the same in the conservation scenario, AB. Note that most developing regions are specified as foreign exchange constrained. Developing regions with rich mineral resource endowments are less likely to be limited by a lack of foreign purchasing power. Their incomes are assumed to be limited by the volume of savings available for investment. The growth rate for the Middle East is set exogenously.

Individual regions are linked by export, import and capital flow equations to simulate the world trade system. Imports of 19 traded goods (minerals trade is discussed below) are each specified as a proportion of total domestic requirements; that proportion, the "import coefficient" for each good, varies from region to region. Values were assigned on the basis of historical trends in specialization. Each region is presumed to export a given proportion of the world pool of total imports (equal to total exports) of each particular commodity. "Export shares" also vary over regions and time to represent regional export specialization.

Most of the coefficients of the system, including import coefficients, export shares, input–output coefficients and consumption proportions are projected to change over time. Thus, the four decadal systems differ because of assumed structural changes that are expected to accompany regional economic development. To a limited extent, successive systems take account of projected changes in technology.

For most commodities in the world system trade is measured on a gross basis, i.e., a region can be both an importer and an exporter of textiles or machinery. For minerals, however, the model explains net, rather than gross, trade. Thus, given exogenous information on resource endowments, regions are classified (for a given decade) as net exporters, as net importers, or as self-sufficient in any given mineral.

Petroleum, natural gas and coal are the major energy resources in the model. The production, for example, of crude petroleum by a net exporter is determined by its own regional requirement for petroleum plus its share of the world pool of demand by net importers. The petroleum extraction capacities of net importers are set exogenously on the basis of information about regional endowments and capacity expected to come on stream. Imports of net importing regions are reckoned as the difference between their total regional demand and regional domestic capacity. Self-sufficient regions are assumed to have enough capacity to satisfy their own domestic needs.

At the time that the model was constructed (1974–77) there was little basis for gauging future conservation and energy consumption coefficients are only marginally lower in the 1990 and 2000 reference systems than they are in the base year. Thus the difference between energy consumption under the selected Cavendish scenario and that in the baseline World Model scenario measures virtually all of the system's response to changing supply conditions over the interval 1980–2000.

I.2. The Cavendish Energy Conservation Scenario

The underlying approach to the design of the Cavendish energy conservation scenario is given in their publications: Bloodworth et al. (1977) and World Energy Conference (1978). These publications describe their methodology and present a variety of scenarios, each based on different assumptions about adaptive response to expected energy supply conditions. The scenario chosen for the present analysis is the most "technologically optimistic" of the range. Called H5, it consists of estimates of energy consumption, specified by region and primary energy source, under the assumption that regional economic growth trends continue at the world average rate of 4.2 per cent per annum through 2020. Given judgment as to energy supplies likely to be available over the period, this scenario represents the amount of energy conservation that would have to be achieved in order to sustain current growth trends. It presumes changes in sectoral techniques and product mix that are as yet unknown but judged to be substantial. The reductions in energy inputs are greater than they would anticipate as normal response to current rates of change in real energy prices; no estimate of the levels of market-clearing energy prices is attempted. The authors do not offer any firm judgement as to whether the energy conservation embodied in the scenario will prove feasible; it is presented as a "conditional projection".

The economic growth rates assumed in the optimistic Cavendish scenario are similar to those in the United Nations Model reference scenario, AA. The H5 scenario was selected because it embodies sufficient conservation to allow the world economy to function at the general levels assumed in the design of the World Model reference scenario. However, energy consumption in the World Model reference scenario is much higher than that of Cavendish H5. If the judgment of the Cavendish group as to future energy availability is correct, the World Model reference scenario will prove infeasible because energy supplies will be inadequate to sustain it. Thus the introduction of the Cavendish scenario might be construed as a "correction" of the baseline World Model projection.

In both the reference and the energy conservation scenario 1980 oil prices were fixed at $21 per barrel and assumed to increase 24 per cent in each of the next two decades. The principal difference between the two scenarios lies not in the prices themselves but in the degree of response to rising prices that is expected. In the baseline scenario conservation responses to rising energy prices were small. Changes in fuel mix and reduced fossil fuel requirements were projected in electric utilities and transportation along with some allowance for synthetic gas from coal. These were introduced my modifying input–output coefficients. Since energy consumption is income-elastic the reference scenario assumptions resulted in a rise in the energy/GDP ratio of roughly eight per cent per decade for the world and three per cent per decade for the average of the developed regions.

In effect the conservation scenario replaces the assumption of very low with that of very high energy price elasticity. Under the energy conservation scenario the world energy/GDP ratio falls by 43 per cent as real energy prices rise 48 per cent between 1980 and 2000. Since the income elasticity of demand for energy is positive, these findings imply a negative price elasticity greater than one. Many economists would consider the assumption of such a large price elasticity overoptimistic but it is difficult to make firm quantitative judgments in this area.

Experience has already shown that the "1980" energy price in the model is too low and many are now revising upward their projections of 1990 and 2000 energy prices. Thus it might be reasonable to expect real price increases of 100 per cent or more over the 1979–2000 interval. Considered against such a price assumption, the conservation scenario would represent a systemwide price response elasticity in the range of -0.4 to -0.5. Most economists would consider such a response plausible.

Before the Cavendish energy conservation scenario could be translated into changes in World Model parameters it was necessary to make some adjustments to improve the consistency of the variables affecting energy consumption in the two systems. In particular, it was necessary to correct for differences in the definitions of regions, for differences in estimates of base year consumption of different fuels and for differences in assumed growth rates of individual regions. These adjustments are described in a mimeographed appendix available on request. Given the special circumstances of the oil producing countries of the Middle East and of Centrally Planned Asia, it seemed expedient to avoid speculation about substantial conservation in these regions. Their projected energy conservation was set at ten per cent of energy consumption in the reference scenario. A summary comparison of energy consumption for other regions in the selected Cavendish and World Model scenarios is presented in Table 2. On average, the Cavendish scenario entails reductions in energy consumption of roughly 35 per cent in non-OPEC developing regions and in the Soviet Union and Eastern Europe and of 30 to 50 per cent in developed market economies.

While the Cavendish work does not publish assumptions as to the real costs of conserving energy, these costs would have a crucial bearing on the economic consequences of conservation measures. Lacking specific information, we introduced an arbitrary assumption that conservation would require additional expenditures on investment and labor equal in value to the current value of energy saved in each region in each decade. The rationale for this lies in the assumption that energy-saving measures will be predominantly "defensive" in the face of rising energy prices: Those conservation measures that would have "paid off" at lower energy prices are already in place. As energy prices rise, more costly arrangements for reducing energy use become economic. Thus, more labor and investment are assumed to be needed per unit of energy saved between 1990 and 2000 than between 1980 and 1990 because the cuts are deeper. A more elegant treatment might have allowed for some portion of the cuts made in the later decade to be made at a cost equal to the 1990 energy price, since some reductions imposed after 1990

Table 2. *Adjusted ratios of Cavendish to world model energy consumption, 1990 and 2000*

	North America	Western Europe	Japan and Oceania	USSR and Eastern Europe	Non-Opec developing
1990					
Oil	.81	.52	.64	.59	.90
Gas	.50	.98	a	.57	.18
Coal	.79	.49	.41	.88	.50
Total	.70	.55	.61	.69	.69
2000					
Oil	.64	.43	.54	.43	.87
Gas	.46	.86	a	.57	.10
Coal	.42	.61	.47	.90	.55
Total	.51	.51	.57	.64	.66

[a] Negligible amounts consumed.

were already economic at 1990 prices. Thus there is some tendency for our assumption to overstate the costs of conservation. On the other hand, some conservation investments may be made in anticipation of future increases in energy prices. This would justify conservation costs in excess of prevailing energy prices.

One could, of course, speculate that technological breakthroughs become more probable with time and that conservation will become progressively cheaper. However, without actually knowing the substance and timing of such technological advances we cannot know their cost advantages nor can we gauge any pessimistic bias of our projections. Meanwhile it may still be overoptimistic to assume that massive conservation will be economic under the assumed price conditions. The costs of energy saving were allocated to investment and labor costs in the ratio of 0.7 to 0.3, by analogy with relative labor and investment costs in the utilities sector. Roughly equal weight was assigned to construction and machinery in conservation capital stock coefficients. Since poorer countries have lower wage rates than richer, the employment intensity of conservation was assumed higher in low than in high income countries.

Had the Cavendish study included projections of changing energy requirements by detailed sectors, it would have been possible to translate their changes into changes in the sectoral energy input coefficients of the World Model. Since, however, it is still too early to know exactly which sectors will achieve particular levels of conservation, and how the product mixes and cost details will change, a shortcut procedure was chosen. The difference between total energy consumption in the baseline and (adjusted) Cavendish scenarios was computed for each region and for each of the three types of energy: coal, petroleum and natural gas. These differences were introduced into a dummy column vector in each region block of the world system as lump sum offsets to the energy demands generated by the coefficients of the original system. Similarly, the volume of additional capital stocks and of annual employment required to effect the stipulated energy savings were calculated, as described above, and introduced as constants in the dummy vector. These constants represent lump sum additions to the capital and labor requirements of the respective regions. The lump sum changes in energy, labor and capital requirements, introduced into the various regional economies, invariably induce some changes in their income (GDP) levels since energy conservation was geared to income growth in the Cavendish scenario. Some iterative adjustments were required to insure consistency between assumed energy conservation and the GDP levels associated with them. However, conservation estimates were conceived as orders of magnitude, rather than precise estimates, and rough adjustments were deemed sufficient.

II.1. Overall Effects on the World Economy

Table 3 summarizes the results of the simulation for the world economy and for each of the three broad groups of regions: Developed (North America, Western Europe (high income), Soviet Union, Eastern Europe, Japan and Oceania), Resource-rich Developing (Middle East, Latin America (low income) and Tropical Africa) and Other Developing (Latin America Medium, Western Europe Medium, Asia Low Income, Arid Africa and Southern Africa). By 2000 the rate of total energy resource extraction in the conservation scenario is lower by roughly 40 per cent as compared with the base scenario. Since conservation was postulated to have a growing effect over the period 1970–2000, the cumulative energy consumption over the period is cut by only 30 per cent. Cumulative outputs of non-energy resources: copper, bauxite, nickel, lead and iron exceed those of the base scenario by percentages varying from about three to seven. Thus conservation involves some substitution of metal for energy resources. Because conservation was assumed to come in the form of new techniques that are more capital intensive than the average in the base scenario, world investment is higher, by almost 12 per cent, than it was in the base runs.

World GDP is slightly higher with conservation than without, largely as a result of the easing of balance of payments constraints in some developing regions (see II.2.2). While increased income means higher consumption in Other Developing regions, on balance, world consumption is lower than it would be without conservation. Higher investment requirements in most regions, and especially in the large developed regions, leave a smaller productive potential available to satisfy consumption. Increased real investment requires increased real savings. The increase in "surplus savings" to a larger negative value means that savings rates, particularly in developed regions, will have to be increased in order to satisfy investment demand.

II.2. Regional Impacts of Conservation

Aggregative as it is, even the summary table (Table 3) indicates that the impact of conservation on GDP, consumption and capital stocks will differ substantially among regions. In terms of GDP, Other Developing regions gain in relation to the Resource-rich and the Developed regions. Inevitably, the findings are dependent on interregional differences in the way income determination is modelled as well as on other differences in regional structure. A summary for the year 2000 of the effects of introducing conservation on individual regions is given in Table 4.

II.2.0. *Conservation and the Middle East*

Conservation as postulated throughout the various regions of the world would reduce Middle Eastern petroleum exports to half their baseline volume in the year 2000 and lower the cumulative petroleum output of this region by 36 per cent. This reduces the total foreign investments of the Middle East to half of their pre-conservation scenario value. Even disregarding any moderating effects on oil prices, their "payments surplus", i.e., additional capital flows required to balance deficits in oil importing regions, would be reduced to less than 40 per cent and foreign investments to half of their respective values in the reference scenario. GDP growth in the Middle East was fixed at 9.5 per cent per year on the assumption that absorptive capacity was limited in that region. When petroleum investment requirements are reduced by the sharp decrease in petroleum exports, the model simulation shows a significant increase in the region's consumption. The volume of imports by the Middle East remains virtually unchanged, despite the decrease in demand for imported investment goods, but the composition of imports shifts to emphasize consumption items, textiles, apparel and food, at the expense of equipment.

II.2.1. *Impact of Conservation in Developed Market Economies*

In general, conservation leads to increased investment at the expense of consumption in developed regions, with minor effects on overall GDP. Given projected labor productivity, the GDP of a developed region is assumed to depend on available labor force. Since the labor supply is fixed *a priori*, increased labor requirements for conservation reduce the amount of labor available for production of other goods and services. Conservation also brings an exogenous increase in capital requirements which, in turn, absorbs some portion of the labor force. On the other hand, reduced energy extraction may

Table 3. *Comparison of energy conservation (AB) and reference (AA) scenarios, 1990 and 2000, by broad classes of regions*[a]

	Scenarios AA and AB Aggregated							
	World				Developed Regions			
	1990AA	1990AB	2000AA	2000AB	1990AA	1900AB	2000AA	2000AB
Macroeconomic indicators								
GDP	7 941	7 971	11 314	11 424	6 308	6 296	8 697	8 688
Consumption	4 914	4 727	7 047	6 872	3 812	3 586	5 319	5 025
Surplus savings	− 96	− 298	− 79	− 324	− 106	− 275	− 51	− 273
Payments surplus					− 346.3	− 148.6	− 750.5	− 263.4
Foreign investments (since 1970)					− 2 501	− 1 334	− 7 838	− 3 222
Investment	1 508	1 717	2 119	2 382	1 278	1 443	1 672	1 892
Equipment	573	732	781	982	459	610	553	729
Plant	930	969	1 333	1 395	818	831	1 117	1 161
Capital stock	16 495	17 557	26 330	28 579	14 095	14 926	21 984	23 767
Equipment	5 141	5 978	7 718	9 356	4 086	4 829	5 749	7 185
Plant	11 354	11 579	18 611	19 222	10 009	10 097	16 234	16 582
Resource extraction								
Copper	17.2	18.8	23.8	25.6	7.0	7.3	9.4	9.8
Bauxite	29.2	30.8	39.9	41.8	9.9	10.3	24.3	25.1
Nickel	1 720	1 871	2 218	2 408	1 066	1 146	1 010	1 060
Zinc	12.8	13.8	17.9	18.9	10.4	11.2	13.9	14.6
Lead	9.9	10.5	14.3	15.0	9.5	10.1	10.5	10.9
Iron	1 151	1 231	1 654	1 740	724	774	1 020	1 068
Petroleum	9 380	6 057	13 774	8 017	3 091	2 200	3 678	2 260
Natural gas	4 146	2 281	5 868	3 160	3 418	1 806	4 652	2 372
Coal	5 576	3 933	9 020	5 972	4 151	2 808	6 882	4 244
Cumulative resource extraction since 1970								
Copper	212.4	222.4	417.7	444.5	101.1	103.6	183.3	189.4
Bauxite	373.3	383.3	718.4	746.1	128.2	130.7	299.3	307.9
Nickel	22 087	22 884	41 779	44 285	13 922	14 376	24 308	25 411
Zinc	167.5	173.6	321.1	337.0	125.5	129.5	247.2	258.6
Lead	123.5	126.4	244.7	254.0	96.4	99.4	196.4	204.6
Iron	14 585	15 068	28 616	29 925	9 840	10 146	15 564	19 360
Petroleum	112 408	93 368	228 184	163 745	41 392	36 967	75 240	59 274
Natural gas	50 819	36 273	100 893	63 485	43 481	33 456	83 836	51 353
Coal	72 882	59 250	145 865	108 778	54 386	43 148	109 557	78 411

[a] Macroeconomic indicators are measured in billions of 1970 dollars except for balance of payments, which reflects relative prices of the forecast year. Metals extraction is measured in millions (nickel in thousands) of metric tons of metal content. Energy is measured in millions of metric tons of coal equivalent.

Resource-rich developing Regions				Other developing Regions			
1990AA	1990AB	2000AA	2000AB	1990AA	1900AB	2000AA	2000AB
408	397	768	754	1 224	1 277	1 848	1 981
236	254	460	507	865	886	1 267	1 339
− 12	− 24	− 74	− 66	22	1	45	15
388.8	191.2	822.0	334.6	− 42.5	− 42.6	− 80.2	− 80.4
3 012	1 858	8 912	4 343	− 511	− 524	− 1 117	− 1 166
82	97	222	212	146	176	224	276
40	44	101	98	74	88	126	154
42	52	120	113	70	85	95	120
587	637	1 561	1 591	1 812	1 993	2 785	3 220
259	272	647	653	795	876	1 321	1 518
327	365	913	938	1 016	1 116	1 463	1 702
6.1	6.9	7.7	8.3	4.1	4.6	6.7	7.4
15.5	16.4	9.0	9.3	3.8	4.1	6.5	7.3
9	11	53	54	643	713	1 154	1 294
1.5	1.6	0.7	0.7	0.9	0.9	3.3	3.6
0.0	0.0	0.9	0.8	0.4	0.4	3.0	3.3
113	122	151	158	313	335	482	512
5 546	3 129	9 253	4 948	742	727	842	808
676	424	1 164	737	51	49	51	51
31	21	46	31	1 393	1 103	2 091	1 696
62.4	66.9	131.8	143.1	48.9	51.9	102.6	112.0
186.9	192.0	309.5	320.5	58.1	60.5	109.6	117.7
247	269	564	597	7 917	8 239	16 906	18 276
21.6	22.6	32.6	34.1	20.5	21.6	41.3	44.4
8.9	8.9	13.4	13.0	18.2	18.1	35.0	36.4
247	1 299	2 576	2 702	3 497	3 622	7 475	7 862
475	46 890	135 477	87 283	9 540	9 510	17 465	17 187
315	4 934	15 523	10 745	1 023	882	1 532	1 387
350	283	739	550	18 144	15 817	35 568	29 816

Table 4. *Selected details from regional solutions, 2000*

	Developed market Regions									
	Middle East		N. Amer.		W. Eur. (high)		Japan		Oceania	
	AA	AB	AA	AB	AA	AB	AA	AB	AA	AB
Per cent growth per annum 1970–2000										
GDP	9.5	9.5	3.4	3.4	3.5	3.5	5.0	5.0	3.9	3.9
Consumption	9.5	10.2	3.2	3.1	3.3	3.1	5.3	5.0	4.2	4.1
Investment	15.0	14.8	4.4	4.6	3.7	4.3	4.2	4.7	4.0	4.3
Equipment	13.3	13.1	3.3	4.1	2.9	4.0	3.1	4.2	3.3	3.7
Construction	14.4	14.2	4.9	5.0	4.3	4.5	4.9	5.0	4.5	4.7
Cap. stock	13.5	13.6	4.4	4.6	4.4	4.8	6.3	6.7	5.1	5.2
Equipment	13.1	13.1	3.5	4.2	3.5	4.5	5.0	6.0	4.0	4.3
Construction	13.8	14.0	4.7	4.7	4.8	4.9	7.0	7.1	5.5	5.6
Billions of 1970 dollars										
Payments surplus[a]	759	287	− 282	− 162	− 298	− 44	− 180	− 84	− 30	− 25
Foreign assets (since 1970)	8 620	4 067	− 2 563	− 1 556	− 3 428	− 832	− 1 897	− 988	− 392	− 351
Surplus savings	− 79	− 72	− 154	− 210	85	27	19	− 5	− 21	− 19
Millions of metric tons of coal equivalent										
Net exports of petroleum	6 042	2 928	− 1 643	− 743	− 2 489	− 914	− 1 401	− 782	− 150	− 79
coal	− 33	− 28	286	84	0	0	− 244	− 37	77	23

[a] Measured in projected relative prices of the year 2000.
[b] Set exogenously.

lead to compensating reductions in labor and capital requirements. However, this offset will take place only if domestic extraction of energy is reduced. To the extent that conservation leads to decreased imports rather than to decreased domestic energy output, it will not tend to lower domestic labor or capital requirements. Since developed market regions are at least partially dependent on imported oil, energy conservation always raises their investment.

Given the labor supply, an increase in investment must be offset by a decrease in consumption. The exact amount of that decrease is determined so as to satisfy the materials, investment and employment balances of the system as a whole. Thus, the net effect of conservation on regional GDP may be positive, negative, or zero, depending on the relative labor content of consumption and investment goods. In summary, the effect of energy conservation on GDP of developed regions depends, first, on the proportions of energy imported and domestic supplies domestically produced. Second, the effect of

Payments constrained Regions														Savings constrained Regions						
L. Amer. (med.)		W. Eur. (med.)		Asia (low)		Arid Af.		South. Af.		L. Amer. (low)		Trop. Af.		Asia Cent. Pl.	
AA	AB	AA	AB	AA	AB	AA	AB	AA	AB	AA	AB	AA	AB	AA	AB
4.4	4.9	4.3	4.8	4.5	4.7	3.0	3.0	4.4	4.5	4.0	3.7	4.5	4.3	3.5	3.5
4.3	4.7	4.3	4.8	4.4	4.5	3.3	3.2	4.4	4.5	3.9	3.7	4.6	4.3	4.7	4.6
4.7	6.0	3.4	4.6	4.5	5.6	3.1	3.7	5.1	4.8	4.9	4.5	4.7	4.7	7.2	7.1
5.6	6.8	4.4	5.4	5.2	6.2	3.7	3.7	5.5	5.5	6.4	6.4	4.7	4.7	7.3	7.2
4.3	5.6	2.3	3.9	5.1	6.3	2.3	3.7	3.7	4.7	2.3	1.3	4.7	4.7	7.1	7.1
5.0	5.9	4.2	5.4	5.6	6.2	3.8	4.1	4.4	4.7	5.5	5.2	6.0	5.9	6.5	6.5
5.3	6.1	4.7	5.7	5.5	6.1	3.5	3.9	4.6	4.8	5.8	5.7	6.0	6.0	6.4	6.4
4.8	5.7	3.9	5.1	5.7	6.2	4.0	4.2	4.3	4.6	5.4	5.0	5.9	5.8	6.6	6.5
b	b	b	b	b	b	b	b	b	b	52	28	12	19	−4	−4
b	b	b	b	b	b	b	b	b	b	222	142	69	132	−57	−59
14	2	19	13	12	1	5	4	9	8	b	b	b	b	b	b
164	−176	−249	−150	−235	−208	0	0	−56	−51	543	263	−74	−60	0	0
−25	−10	−78	−36	4	1	−1	0	0	0	−6	−2	0	0	19	6

increased investment requirements on GDP depends on the total domestic labor content of consumption as compared with investment goods.

In the present simulation, energy conservation led to trivial decreases in GDP for the three largest developed regions, North America, Western Europe (high income) and Japan and to a tiny increase for Oceania. All of these regions require larger capital stocks and higher rates of investment and must make some consumption sacrifice in order to accomplish the stipulated conservation.

In all developed market regions, total equipment stock is affected more markedly than construction when energy conservation is introduced (Table 4). This may be surprising, since it was assumed that the capital costs of conservation were split roughly equally between equipment and construction. However, bear in mind that conservation leads to some decreases in consumption levels. Thus, increases in construction requirements directly associated with conservation are offset by direct and indirect decreases in construction (primarily housing) that go with reduced consumption. In effect, the activity "energy saving" is more equipment-intensive but barely more construction-intensive than the average of all activities in the economy's product mix in affluent regions. In low income regions, where housing capital coefficients are much smaller than in developed market regions, energy conservation leads to significant increases in both construction and equipment capital stocks.

Increased investment requirements imply an increased demand for savings when energy conservation is introduced. In North America, with conservation, savings are estimated to fall roughly 30 per cent below the volume necessary to sustain required investment in 2000. While projections of savings rates are crude at best, this finding suggests that savings would have to be somewhat larger than those presently considered "normal" at projected levels of income per capita. An increase in the savings rate from the presently assumed 17 to 23 per cent would be necessary to close the gap.

Differences in the effects of energy conservation in individual regions in this group reflect some differences in the postulated energy conservation programs as well as differences in the regional economies. While overall energy conservation was of similar magnitude for all developed regions, it took the form of large cutbacks in natural gas in North America, where supplies are becoming tight, and only minor cutbacks in Europe, where new supplies are presumed to come on line. To some extent, gas is presumed to substitute for petroleum in Western Europe, permitting relatively larger conservation of petroleum in Europe than in other regions.

Among the developed market regions, the net effect on capital requirements of the conservation scenario is smallest in North America. In Europe, capital costs of conservation are augmented by the capital costs of shifting from imported oil to domestically produced natural gas. In North America, the relative contribution of domestically produced gas decreases. Both of these regions extract significant amounts of energy domestically and capital costs of conservation are partially offset by reductions in capital costs of extraction when energy extraction is reduced. For Japan, where virtually all energy is imported, there is no offset to the capital requirements for effecting conservation.

The major benefit of conservation to developed regions is an appreciable reduction of balance of payments deficits. The improvement is most marked for Western Europe (high income), where the postulated curtailment of energy consumption leads to a condition that eliminates the payments deficits projected in the base scenario. Since North America produces a significant proportion of its energy requirements domestically, one might expect energy conservation to bring a larger relative improvement to its payments position than to that of Japan, who imports virtually all of her energy. If we look at petroleum alone, this is indeed the case. However, coal trade complicates the picture. Japan is a net importer of coal and energy conservation reduces her coal imports substantially. North America, on the other hand, is expected to become a substantial net exporter of coal by the year 2000. A worldwide energy conservation scenario reduces the demand for North American coal exports at the same time that it reduces North American petroleum imports. Both of these tendencies are reflected in the balance of payments totals. In addition, petroleum imports comprise a larger proportion of the import bill in Japan than they do in North America.

II.2.2. *Impact of Conservation in Payments-Constrained Regions*

In the reference scenario, most developing regions are specified as payments-constrained. In general, energy conservation tends to increase the incomes of these regions by reducing their energy imports. Reductions of the energy import bill are partially offset by increases in requirements for imported investment goods necessary for conservation. Since investment requirements increase, consumption tends to rise somewhat less than overall GDP when energy conservation is implemented.

Different economic characteristics of the regions explain differential impacts of conservation. The income augmenting effect of energy conservation is greatest for the two "middle income" developing regions, Latin America (medium income) and Western Europe (medium income). Because they are relatively industrialized and prosperous, energy constitutes a larger proportion of their total import bills than in most other regions of their class. Hence the effects of conservation are more pronounced. Southern Africa is assumed self-sufficient in coal and hence conservation does not reduce her import bill as much as those of other middle-income regions.

Arid Africa is the only one of the payments-constrained regions whose GDP falls (slightly) with the conservation scenario. Since this region is modelled as self-sufficient in petroleum, energy conservation does not reduce its import requirements. Energy conservation is assumed to be more capital intensive than oil extraction in this region. Since the bulk of this region's capital equipment is imported, the conservation program results in an increase in import requirements and a reduction in equilibrium income level.

II.2.3. *Impact of Conservation in Savings-Constrained Regions*

The World Model has fixed savings rates for each region and decade. As income rises, investment requirements tend to increase more rapidly than savings. Hence, a rise in investment intensity tends to lower equilibrium GDP. There are only a few regions specified as savings-constrained and each is in some sense a special case. Tropical Africa is the most straightforward. It was classed as savings- rather than payments-constrained because it is a net exporter of non-fuel minerals and these provide a source of foreign exchange. Energy conservation increases this region's investment requirements, reducing its GDP and consumption slightly. The balance of payments surplus of this region increases as its net imports of petroleum and other products fall.

Latin America (low income) is a net exporter of petroleum. Energy conservation brings increased investment requirements and lower GDP. The balance of payments shows the net effect of two opposite influences: reduction in import requirements and increased exports of non-fuel minerals tend to improve the payments position of the region; reductions in petroleum exports due to reduced demand by other regions tend to deteriorate it. The second influence dominates in 1990; the first dominates in 2000.

Centrally Planned Asia is modelled as self-sufficient in petroleum and natural gas and as a net exporter of coal. Energy conservation, though modest by assumption, exerts downward pressure on GDP and, more markedly, on consumption in this region. Because other regions' demands for coal imports decline, this region's balance of payments deteriorates slightly.

III. Conclusions

This study began with a very vigorous energy conservation scenario that would permit world economic development to continue through 2000 without exceeding the constraints of probable energy supply levels. Analyzing this conservation scenario in the context of the United Nations World Model gives some basis for appraising the economic effects of the conservation program. In general, the results indicate that the energy conservation scenario alleviates some of the important stresses initiated by the mid-seventies oil price increase but does not eliminate them. The payments surplus of the Middle East would be reduced to about half of its value in the reference scenario. To the extent that this region has limited absorptive capacity, the remaining surplus remains to be offset through asset transfers ("recycling") or, less likely, through a reduction in petroleum prices.

Conservation itself is bound to entail costs. Under the optimistic cost assumptions of this study, a 30 per cent reduction in cumulative energy use by the year 2000 would be achieved through an increase of 8.5 per cent in gross capital stock. Investment rates would have to increase by 12.4 per cent to offset the decrease of 40 per cent in the annual rate of energy consumption, by the year 2000. Personal consumption levels would be reduced, correspondingly, by 2.5 per cent and increased rates of saving would be necessary.

While conservation would substantially reduce pressures in the world payments system, regional implications vary greatly. Net exporters of energy would have smaller payments surpluses and foreign asset balances. On the other hand, their domestic stocks of nonrenewable energy resources would be depleted less rapidly. Essentially they would trade domestic for foreign assets. To the extent that conservation increases the viability of other regional economies it protects the value of the Middle East's foreign assets. Thus it is not clear that the position of petroleum exporters would be deteriorated unequivocally by conservation.

The balance of payments positions of developed market economies would be substantially improved at the expense of minor reductions in consumption levels. Payments-constrained developing regions would be relieved of part of their energy import costs, although some of the reductions would be offset by added expenditures for imported equipment. Regions whose primary constraint is the supply of savings for investment purposes would experience increased difficulties when faced with the higher investment requirements of the conservation scenario. The benefits of conservation for this last group are not really clear.

In interpreting the results of these computations it is important to remember that the details of conservation techniques were not fully articulated and that price responses are exogenous in the World Model. Thus, nothing has been said about the effects of conservation in moderating pressures toward future increases in world energy prices, although that might indeed turn out to be a crucial issue in international economic relations. The conservation scenario did not indicate the prices at which the postulated conservation might be expected to be accomplished, should that conservation prove feasible at all.

Since it does not allow for endogenous price responses, the World Model underestimates the possible adjustments that individual regions can make to balance of payments difficulties. Recent history has shown some regions much more successful than others in expanding their manufactured exports to compensate for increased costs of imported energy. In its present form, the model does not subsume exchange rate adjustments and their effects on export shares.

On the other hand, the model does highlight some realistic limitations of the adjustment mechanism. Unless the regional distribution of energy endowments is itself modified, the success of a few regions in improving their balances by expanding exports cannot be generalized to all. So long as the Middle East has a favorable balance of trade, the rest of the world's combined balance must be unfavorable. If Japan, Brazil, Korea, West Germany, etc., improve their positions through increased exports of manufactured goods, this is done at the expense of the share of other industrial producers. Only reduced oil prices, reduced oil imports or increased imports of goods by oil exporting regions can eliminate this fundamental problem.

The general conclusions of the study remain sensitive to the key assumptions about the course of real energy prices and prospective costs of conservation. Projections of real energy prices are probably too low. Increasing them would tend to make the conservation scenario more plausible but it would certainly worsen the projected imbalances in the world payments system. Lower-cost conservation opportunities would lower the consumption sacrifice required in developed regions and brighten the growth prospects of the developing world. We all hope that the requisite technology will materialize.

References

Bloodworth, I. J., Bossanyi, E., Bowers, D. W., Crouch, E. A. C., Eden, R. J., Hope, C. W., Humphrey, W. S., Mitchell, J. V., Pullin, D. J. & Stanislaw, J. A.: *World Energy Demand to 2020*. Executive Summary. Published by the World Energy Conference, London, 1977.

Carter, Anne P. & Petri, Peter A.: Factors Affecting the Long Term Prospects of Developing Regions. *Journal of Policy Modeling 1* (3), 359–381, 1979.

Leontief, W. W., Carter, A. P. & Petri, P. A.: *The Future of the World Economy*. Oxford University Press, New York, 1977.

World Energy Conference, *World Energy Resources, 1985–2020*. IPC Science and Technology Press, London and New York Technology Press, London and New York, 1978.

ENERGY PRICES AND PRODUCTIVITY GROWTH

Dale W. Jorgenson

Harvard University, Cambridge, Massachusetts, USA

Abstract

The sharp decline in economic growth since 1973 presents a problem comparable in scientific interest and social importance to the problem of unemployment in the Great Depression of the 1930s. Conventional methods of economic analysis have been tried and have been found to be inadequate. This paper outlines a new framework for understanding the slowdown in economic growth. We find that the fall in U.S. economic growth since 1973 has been due to the dramatic decline in productivity growth. Within the new framework we identify higher energy prices as an important determinant of the productivity slowdown.

I. Introduction

The growth of the U.S. economy in the postwar period has been very rapid by historical standards. The rate of economic growth reached its maximum during the period 1960 to 1966. Growth rates have slowed substantially since 1966 and declined further since 1973. A major source of uncertainty in projections of the future of the U.S. economy is whether patterns of growth will better conform to the rapid growth of the early 1960s, the more moderate growth of the late 1960s and early 1970s or the disappointing growth since 1973.

In this paper our first objective is to identify the sources of uncertainty about future U.S. economic growth more precisely. For this purpose we decompose the growth of output during the postwar period into contributions of capital input, labor input, and productivity growth. For the period 1948 to 1976 we find that all three sources of economic growth are significant and must be considered in analyzing future growth potential. For the postwar period capital input has made the most important contribution to the growth of output, productivity growth has been next most important, and labor input has been least important.

Focusing on the period 1973 to 1976, we find that the fall in the rate of economic growth has been due to a dramatic decline in productivity growth. Declines in the contributions of capital and labor input are much less significant in explaining the slowdown. We conclude that the future growth of productivity is the main source of uncertainty in projections of future U.S. economic growth.

Our second objective is to analyze the slowdown in productivity growth for the U.S. economy as a whole in greater detail. For this purpose we decompose productivity growth during the postwar period into components that can be identified with productivity growth at the sectoral level and with reallocations of output, capital input, and labor input among sectors. For the period 1948 to 1976, we find that these reallocations are insignificant relative to sectoral productivity growth. The combined effect of all three reallocations is slightly negative, but sufficiently small in magnitude to be negligible as a source of aggregate productivity growth.

Again focusing on the period 1973 to 1976, it is possible that the economic dislocations that accompanied the severe economic contraction of 1974 and 1975 could have resulted in shifts of output and inputs among sectors that contributed to the slowdown of productivity growth at the aggregate level. Alternatively, the sources of the slowdown might be found in slowing productivity growth at the level of individual industrial sectors. We find that the contribution of reallocations of output and inputs among sectors was positive rather than negative during the period 1973–1976 and relatively small. Declines in productivity growth for the individual industrial sectors of the U.S. economy must bear the full burden of explaining the slowdown in productivity growth for the economy as a whole.

The decomposition of the growth of output among contributions of capital input, labor input, and productivity growth is helpful in isolating the sources of uncertainty in future growth projections. The further decomposition of productivity growth among the reallocations of output, capital input, and labor input among sectors and growth in productivity at the sectoral level provides additional detail. The uncertainty in future growth projections can be resolved only by providing an explanation for the fall in productivity growth at the sectoral level. For this purpose an econometric model of sectoral productivity growth is required.

Our third objective is to present the results of an econometric analysis of the determinants of productivity growth at the sectoral level. Our econometric model determines the growth of sectoral productivity as a function of relative prices of sectoral inputs. For each sector we divide inputs among capital, labor, energy, and materials inputs. We allow for the fact that the value of sectoral output includes the value of intermediate inputs—energy and materials—as well as the value of primary factors of production—capital and labor. Differences in relative prices for inputs are associated with differences in productivity growth for each sector.

After fitting our econometric model of productivity growth to data for individual industrial sectors we find that productivity growth decreases with an increase in the price of capital input for a very large proportion of U.S. industries. Similarly, productivity growth falls with higher prices of labor input for a large proportion of industries. The impact of higher energy prices is also to slow the growth of productivity for a large proportion of industries. By contrast we find that an increase in the price of materials input is associated with increases in productivity growth for almost all industries.

Since 1973 the relative prices of capital, labor, energy, and materials inputs have been altered radically as a consequence of the increase in the price of energy relative to other productive inputs. Higher world petroleum prices following the Arab oil embargo of late 1973 and 1974 have resulted in sharp increases in prices for all forms of energy in the U.S. economy—oil, natural gas, coal, and electricity generated from fossil fuels and other sources. Although the U.S. economy has been partly shielded from the impact of higher world petroleum prices through a system of price controls, all industrial sectors have experienced large increases in the price of energy relative to other inputs.

Our econometric model reveals that slower productivity growth at the sectoral level is associated with higher prices of energy relative to other inputs. Our first conclusion is that the slowdown of sectoral productivity growth after 1973 is partly a consequence of the sharp increase in the price of energy relative to other productive inputs that began with the run-up of world petroleum prices in late 1973 and early 1974. The fall in sectoral productivity growth after 1973 is responsible in turn for the decline in productivity growth for the U.S. economy as a whole. Slower productivity growth is the primary source of the slowdown in U.S. economic growth since 1973.

Our final objective is to consider the prospects for future U.S. economic growth. Exports of petroleum from Iran dropped sharply during 1979, following the revolution in that country in late 1978. During 1979 world petroleum prices have jumped 130 to 140 per cent, resulting in large and rapid price increases for petroleum products in the U.S. During 1979 the prices of petroleum products began to move to world levels as a consequence of the decontrol of domestic prices by the U.S. government over the period 1979 to 1981. Prices of natural gas will also be allowed to rise through decontrol by 1985 or, at the latest, by 1987. Prices of energy confronted by individual industries within the United States have already increased relative to other productive inputs and can be expected to increase further.

Based on the performance of the U.S. economy since 1973, we can anticipate a further slowdown in the rate of economic growth, a decline in the growth of productivity for the economy as a whole, and declines in sectoral productivity growth for a wide range of industries. These dismal conclusions suggest that a return to rapid growth of the early 1960's is highly unlikely, that even the slower growth of the late 1960's and early 1970's will be difficult to attain, and that the performance of the U.S. economy during the 1980s could be worse than during the period from 1973 to the present. We conclude the paper

Table 1. *Growth of output and inputs for the U.S. economy, 1948–1976*

	1948–1976	1948–1953	1953–1957	1957–1960	1960–1966	1966–1969	1969–1973	1973–1976
Growth rates								
Output	0.0350	0.0457	0.0313	0.0279	0.0483	0.0324	0.0324	0.0089
Capital input	0.0401	0.0507	0.0393	0.0274	0.0376	0.0506	0.0396	0.0312
Labor input	0.0128	0.0160	0.0023	0.0099	0.0199	0.0185	0.0166	0.0058
Productivity	0.0114	0.0166	0.0146	0.0113	0.0211	0.0004	0.0095	−0.0070
Contributions								
Capital input	0.0161	0.0194	0.0154	0.0109	0.0156	0.0211	0.0161	0.0126
Labor input	0.0075	0.0097	0.0013	0.0057	0.0116	0.0108	0.0068	0.0033

with a discussion of policy measures to ameliorate the negative effects of higher energy prices on future U.S. economic growth.

II. The Growth Slowdown

In this section we begin our analysis of the slowdown in U.S. economic growth by decomposing the growth of output for the economy as a whole into the contributions of capital input, labor input, and productivity growth.[1] The results are given in Table 1 for the postwar period 1948–1976 and for the following seven subperiods—1948–1953, 1953–1957, 1957–1960, 1960–1966, 1966–1969, 1969–1973, and 1973–1976.[2] Except for the period from 1973 to 1976, each of the subperiods covers economic activity from one cyclical peak to the next. The last period covers economic activity from the cyclical peak in 1973 to 1976, a year of recovery from the sharp downturn in economic activity in 1974 and 1975.

We first present rates of growth for output, capital input, labor input, and productivity for the U.S. economy. For the postwar period as a whole output grew at 3.50 per cent per year, capital input grew at 4.01 per cent, and labor grew at 1.28 per cent. The growth of productivity averaged 1.14 per cent per year. The rate of economic growth reached its maximum at 4.83 per cent during the period 1960–1966 and grew at only 0.89 per cent during the recession and partial recovery of 1973–1976. The growth of capital input was more even, exceeding 5 per cent in 1948–1953 and 1966–1969 and falling to 3.12 per cent in 1973–1976. The growth of labor input reached its maximum in the period 1960–1966 at 1.99 per cent and fell to 0.58 per cent in 1973–1976, which was above the minimum of 0.23 per cent in the period 1953–1957.

---

[1] The methodology that underlies our decomposition of the growth of output is presented in detail by Jorgenson (1980).
[2] The results presented in Table 1 are those of Fraumeni & Jorgenson (1980), who also provide annual data for output and inputs.

We can express the rate of growth of output for the U.S. economy as a whole as the sum of a weighted average of the rates of growth of capital and labor inputs and the growth of productivity. The weights associated with capital and labor inputs are average shares of these inputs in the value of output. The contribution of each input is the product of the average share of this input and corresponding input growth rate. We present contributions of capital and labor inputs to U.S. economic growth for the period 1948–1976 and for seven sub-periods in Table 1. Considering productivity growth, we find that the maximum occurred from 1960 to 1966 at 2.11 per cent per year. During the period 1966–1969 productivity growth was almost negligible at 0.04 per cent. Productivity growth recovered to 0.95 per cent during the period 1969–1973 and fell to a negative 0.70 per cent during 1973–1976.

Since the value shares of capital and labor inputs are very stable over the period 1948–1976, the movements of the contributions of these inputs to the growth of output largely parallel those of the growth rates of the inputs themselves. For the postwar period as a whole the contribution of capital input of 1.61 per cent is the most important source of output growth. Productivity growth is next most important at 1.14 per cent, while the contribution of labor input is the third most important at 0.75 per cent. All three sources of growth are significant and must be considered in an analysis of the slowdown of economic growth during the period 1973–1976. However, capital input is clearly the most important contributor to the rapid growth of the U.S. economy during the postwar period.[1]

Focusing on the period 1973 to 1976, we find that the contribution of capital input fell to 1.26 per cent for a drop of 0.35 per cent from the postwar average, the contribution of labor input fell to 0.33 per cent for a drop of 0.42 per cent, and that productivity growth at a negative 0.70 per cent dropped 1.84 per cent. We conclude that the fall in the rate of U.S. economic growth during the period 1973–1976 was largely due to the fall in productivity growth. Declines in the contributions of capital and labor inputs are much less significant in explaining the slowdown. A detailed explanation of the fall in productivity growth is needed to account for the slowdown in U.S. economic growth.

To analyze the sharp decline in productivity growth for the U.S. economy as a whole during the period 1973 to 1976 in greater detail we employ data on productivity growth for individual industrial sectors. For this purpose it is important to distinguish between productivity growth at the aggregate level and productivity growth at the sectoral level. At the aggregate level the appropriate concept of output is value added, defined as the sum of the values of capital and labor inputs for all sectors of the economy. At the sectoral level the appropriate concept of output includes the value of primary factors of production at the sectoral level—capital and labor inputs—and the value of

---

[1] This conclusion contrasts sharply with that of Denison (1979). For a comparison of our methodology with that of Denison, see Jorgenson & Griliches (1972).

Table 2. *Productivity growth for the U.S. economy 1948–1976*

	1948– 1976	1948– 1953	1953– 1957	1957– 1960	1960– 1966	1966– 1969	1969– 1973	1973– 1976
Sectoral productivity growth	0.0124	0.0219	0.0177	0.0145	0.0217	0.0025	0.0048	−0.0113
Reallocation of value added	−0.0016	−0.0075	−0.0030	−0.0010	−0.0016	−0.0025	0.0030	0.0046
Reallocation of capital input	0.0008	0.0022	0.0008	−0.0001	0.0002	0.0001	0.0010	0.0008
Reallocation of labor input	−0.0002	−0.0000	−0.0008	−0.0021	0.0008	0.0004	0.0006	−0.0011
Productivity	0.0114	0.0166	0.0146	0.0113	0.0211	0.0004	0.0095	−0.0070

intermediate inputs—energy and materials inputs. In aggregating over sectors to obtain output for the U.S. economy as a whole the production and consumption of intermediate goods cancel out, so that values of energy and materials inputs do not appear at the aggregate level.

We can express productivity growth for the U.S. economy as a whole as the sum of four components. The first component is a weighted sum of productivity growth rates for individual industrial sectors. The weights are ratios of the value of output in each sector to value added in that sector. The sum of these weights over all sectors exceeds unity, since productivity growth in each sector contributes to the growth of output in that sector and to the growth of output in other sectors through deliveries of intermediate inputs to those sectors. The remaining components of aggregate productivity growth represent the contributions of reallocations of value added, capital input, and labor input among sectors to productivity growth for the economy as a whole.[1]

The role of reallocations of output, capital input and labor input among sectors is easily understood. For example, if capital input moves from a sector with a relatively low rate of return to a sector with a high rate of return, the quantity of capital input for the economy as a whole is unchanged, but the level of output is increased, so that productivity has improved. Similarly, if labor input moves from a sector with low wages to a sector with high wages, labor input is unchanged, but productivity has improved. Productivity growth for the economy as a whole is the sum of productivity growth at the sectoral has and reallocations of output, capital input and labor input among sectors. Data on sectoral productivity growth, reallocations of output, capital input, and labor input, and aggregate productivity growth for the postwar period 1948 to 1976 and for seven subperiods are given in Table 2.[2]

[1] The methodology that underlies our decomposition of productivity growth is presented in detail by Jorgenson (1980).
[2] The results presented in Table 2 are those of Fraumeni & Jorgenson (1980), who also provide annual data for productivity growth.

Productivity growth at the aggregate level is dominated by the contribution of sectoral productivity growth of 1.24 per cent per year for the postwar period as a whole. The contributions of reallocations of output, capital input, and labor input are a negative 0.16 per cent, a positive 0.08 per cent, and a negative 0.02 per cent. Adding these contributions together we find that the combined effect of the three reallocations is a negative 0.10 per cent, which is negligible by comparison with the effect of productivity growth at the sectoral level. Productivity growth at the aggregate level provides an accurate picture of average productivity growth for individual industries; this picture is not distorted in an important way by the effect of reallocations of output and inputs among sectors.

Again focusing on the period 1973–1976, we find that the contribution of sectoral productivity growth to productivity growth for the economy as a whole fell to a negative 1.13 per cent for a drop of 2.37 per cent from the postwar average. By contrast the contribution of reallocations of output rose to 0.46 per cent for a gain of 0.62 per cent from the postwar average. The contribution of the reallocation of capital input was unchanged at 0.08 per cent, while the contribution of labor input fell to a negative 0.11 per cent for a drop of 0.09 per cent from of the postwar average. The combined contribution of all three reallocations rose 0.53 per cent, partially offsetting the precipitous decline in productivity growth at the sectoral level. We conclude that declines in productivity growth for the individual industrial sectors of the U.S. economy are more than sufficient to explain the decline in productivity growth for the economy as a whole.

To summarize our findings on the slowdown of U.S. economic growth during the period 1973–1976: We find that the drop in the growth of output of 2.61 per cent per year from the postwar average is the sum of a decline in the contribution of labor input of 0.42 per cent per year, a sharp dip in sectoral rates of productivity growth of 2.37 per cent, a rise in the role of reallocations of output among sectors of 0.62 per cent per year, no change in the reallocations of capital input, and a decline in the contribution of reallocations of labor input of 0.09 per cent per year. Whatever the causes of the slowdown, they are to be found in the collapse of productivity growth at the sectoral level rather than a slowdown in the growth of capital and labor inputs at the aggregate level or the reallocations of output, capital input, or labor input among sectors.

The decomposition of economic growth into the contributions of capital input, labor input, and productivity growth is helpful in pinpointing the causes of the slowdown. The further decomposition of productivity growth for the economy as a whole into contributions of sectoral productivity growth and reallocations of output, capital input, and labor input is useful in providing additional detail. However, our measure of sectoral productivity growth is simply the unexplained residual between growth of sectoral output and the contributions of sectoral capital, labor, energy, and materials inputs. The problem remains of providing an explanation for the fall in productivity growth at the sectoral level.

III. Sectoral Productivity Growth

We have now succeeded in identifying the decline in productivity growth at the level of individual industrial sectors within the U.S. economy as the main culprit in the slowdown of U.S. economic growth that took place after 1973. To provide an explanation for the slowdown we must go behind the measurements to identify the determinants of productivity growth at the sectoral level. For this purpose we require an econometric model of sectoral productivity growth. In this section we present a summary of the results of applying such an econometric model to detailed data on sectoral output and capital, labor, energy, and materials inputs for thirty-five individual industries in the United States.

Our complete econometric model is based on sectoral price functions for each of the thirty-five industries included in our study.[1] Each price function give the price of the output of the corresponding industrial sector as a function of the prices of capital, labor, energy, and materials inputs and time, where time represents the level of technology in the sector.[2] Obviously, an increase in the price of one of the inputs, holding the prices of the other inputs and the level of technology constant, will necessitate an increase in the price of output. Similarly, if productivity in a sector improves and the prices of all inputs into the sector remain the same, the price of output must fall. Price functions summarize these and other relationships among the prices of output, capital, labor, energy, and materials inputs, and the level of technology.

Although the sectoral price functions provide a complete model of production patterns for each sector, it is useful to express this model in an alternative and equivalent form. We can express the shares of each of the four inputs—capital, labor, energy, and materials—in the value of output as functions of the prices of these inputs and time, again representing the level of technology.[3] We can add to these four equations for the value shares an equation that expresses productivity growth as a function of the prices of the four inputs and time.[4] In fact, the negative of the rate of productivity growth is a function of

---

[1] Econometric models for each of the thirty-five industries are given by Jorgenson & Fraumeni (1981).

[2] The price function was introduced by Samuelson (1953). A complete characterization of the sectoral price functions employed in this study is provided by Jorgenson & Fraumeni (1981).

[3] Our sectoral price functions are based on the translog price function introduced by Christensen, Jorgenson & Lau (1971, 1973). The translog price function was first applied at the sectoral level by Berndt & Jorgenson (1973) and Berndt & Wood (1975). References to sectoral production studies incorporating energy and materials inputs are given by Berndt & Wood (1979).

[4] Productivity growth is represented by the translog index introduced by Christensen & Jorgenson (1970). The translog index of productivity growth was first derived from the translog price function by Diewert (1980) and by Jorgenson & Lau (1981).

the four input prices and time. This equation is our econometric model of sectoral productivity growth.[1]

Like any econometric model, the relationships determining the value shares of capital, labor, energy, and materials inputs and the negative of the rate of productivity growth involve unknown parameters that must be estimated from data for the individual industries. Included among these unknown parameters are biases of productivity growth that indicate the effect of changes in the level of technology on the value shares of each of the four inputs.[2] For example, the bias of productivity growth for capital input gives the change in the share of capital input in the value of output in response to changes in the level of technology, represented by time. We say that productivity growth is capital using if the bias of productivity growth for capital input is positive. Similarly, we say that productivity growth is capital saving if the bias of productivity growth for capital input is negative. The sum of the biases for all four inputs must be precisely zero, since the changes in all four shares with any change in technology must sum to zero.

We have pointed out that our econometric model for each industrial sector of the U.S. economy includes an equation giving the negative of sectoral productivity growth as a function of the prices of the four inputs and time. The biases of technical change with respect to each of the four inputs appear as the coefficients of time, representing the level of technology, in the four equations for the value shares of all four inputs. The biases also appear as coefficients of the prices in the equation for the negative of sectoral productivity growth. This feature of our econometric model makes it possible to use information about changes in the value shares with time and changes in the rate of sectoral productivity growth with prices in determining estimates of the biases of technical change.

The biases of productivity growth express the dependence of value shares of the four inputs on the level of technology and also express the dependence of the negative of productivity growth on the input prices. We can say that capital using productivity growth, associated with a positive bias of productivity growth for capital input, implies that an increase in the price of capital input decreases the rate of productivity growth (or increases the negative of the rate of productivity growth). Similarly, capital saving productivity growth, associated with a negative bias for capital input, implies that an increase in the price of capital input increases the rate of productivity growth. Analogous relationships hold between biases of labor, energy, and materials inputs and

---

[1] This model of sectoral productivity growth is based on that of Jorgenson & Lau (1981).
[2] The bias of productivity growth was introduced by Hicks (1932). An alternative definition of the bias of productivity growth was introduced by Binswanger (1974a, 1974b). The definition of the bias of productivity growth employed in our econometric model is due to Jorgenson & Lau (1981).

the direction of the impact of changes in the prices of each of these inputs on the rate of productivity growth.[1]

Jorgenson & Fraumeni (1981) have fitted biases of productivity growth for thirty-five industrial sectors that make up the whole of the producing sector of the U.S. economy. They have also fitted the other parameters of the econometric model that we have described above. Since our primary concern in this section is to analyze the determinants of productivity growth at the sectoral level, we focus on the patterns of productivity growth revealed in Table 3. We have listed the industries characterized by each of the possible combinations of biases of productivity growth, consisting of one or more positive biases and one or more negative biases.[2]

The pattern of productivity growth that occurs most frequently in Table 3 is capital using, labor using, energy using, and materials saving productivity growth. This pattern occurs for nineteen of the thirty-five industries analyzed by Jorgenson and Fraumeni. For this pattern of productivity growth the bias of productivity growth for capital input, labor input, and energy input are positive, and the bias of productivity growth for materials input is negative. This pattern implies that increases in the prices of capital input, labor input, and energy input decrease the rate of productivity growth, while increases in the price of materials input increase the rate of productivity growth.

Considering all patterns of productivity growth included in Table 3, we find that productivity growth is capital using for twenty-five of the thirty-five industries included in our study. Productivity growth is capital saving for the remaining ten industries. Similarly, productivity growth is labor using for thirty-one of the thirty-five industries and labor saving for the remaining four industries; productivity growth is energy using for twenty-nine of the thirty-five industries included in Table 3 and is energy saving for the remaining six. Finally, productivity growth is materials using for only two of the thirty-five industries and is materials saving for the remaining thirty-three. We conclude that for a very large proportion of industries the rate of productivity growth decreases with increases in the prices of capital, labor, and energy inputs, and increases in the price of materials inputs.

The most striking change in the relative prices of capital, labor, energy and materials inputs that has taken place since 1973 is the staggering increase in the price of energy. The rise in energy prices began in 1972 before the Arab oil embargo, as the U.S. economy moved toward the double-digit inflation that characterized 1973. In late 1973 and early 1974 the price of petroleum on world markets increased by a factor of four, precipitating a rise in domestic prices of petroleum products, natural gas, coal, and uranium. The impact of higher

---

[1] A complete characterization of biases of productivity growth is given by Jorgenson & Fraumeni (1981).

[2] The results presented in Table 3 are those of Jorgenson & Fraumeni (1981). Of the fourteen logically possible combinations of biases of productivity growth, only the eight patterns presented in Table 3 occur empirically.

Table 3. *Classification of industries by biases of productivity growth*

Pattern of biases	Industries
Capital using Labor using Energy using Material saving	Agriculture, metal mining, crude petroleum and natural gas, nonmetallic mining, textiles, apparel, lumber, furniture, printing, leather, fabricated metals, electrical machinery, motor vehicles, instruments, miscellaneous manufacturing, transportation, trade, finance, insurance and real estate, services
Capital using Labor using Energy saving Material saving	Coal mining, tobacco manufactures, communications, government enterprises
Capital using Labor saving Energy using Material saving	Petroleum refining
Capital using Labor saving Energy saving Material using	Construction
Capital saving Labor saving Energy using Material saving	Electric utilities
Capital saving Labor using Energy saving Material saving	Primary metals
Capital saving Labor using Energy using Material saving	Paper, chemicals, rubber, stone, clay and glass, machinery except electrical, transportation equipment and ordnance, gas utilities
Capital saving Labor saving Energy using Material using	Food

world petroleum prices was partly deflected by price controls for petroleum and natural gas that resulted in the emergence of shortages of these products during 1974. All industrial sectors of the U.S. economy experienced sharp increases in the price of energy relative to other inputs.

Slower growth in productivity at the sectoral level is associated with higher energy prices for twenty-nine of the thirty-five industries that make up the producing sector of the U.S. economy. The dramatic increases in energy prices resulted in a slowdown in productivity growth at the sectoral level. In the preceding section we have seen that the fall in sectoral productivity growth after

1973 is the primary explanation for the decline in productivity for the U.S. economy as a whole. Finally, we have shown that the slowdown in productivity growth during the period 1973–1976 is the main source of the fall in the rate of U.S. economic growth since 1973.

We have now provided part of the solution to the problem by the disappointing growth record of the U.S. economy since 1973. By reversing historical trends toward lower prices of energy in the U.S. economy, the aftermath of the Arab Oil Embargo of 1973 and 1974 has led to an end to rapid economic growth. The remaining task is to draw the implications of our findings for future U.S. economic growth. Projections of future economic growth must take into account the dismal performance of the U.S. economy since 1973 as well as the rapid growth that has characterized the U.S. economy during the postwar period. In particular, such projections must take into account the change in the price of energy input for individual industrial sectors, relative to prices of capital, labor, and materials inputs.

IV. Prognosis

Our objective in this concluding section of the paper is to provide a prognosis for future U.S. economic growth. For this purpose we cannot rely on the extrapolation of past trends in productivity growth or its components. The year 1973 marks a sharp break in trend associated with a decline in rates of productivity growth at the sectoral level. Comparing the period after 1973 with the rest of the postwar period, we can associate the decline in productivity growth with the dramatic increase in energy prices that followed the Arab oil embargo in late 1973 and early 1974. The remaining task is to analyze the prospects for a return to the high sectoral productivity growth rates of the early 1960s, for moderate growth of sectoral productivity growth like that of the late 1960s and early 1970s, or for continuation of the disappointing growth since 1973.

During 1979 there has been a further sharp increase in world petroleum prices, following the interruption of Iranian petroleum exports that accompanied the revolution that took place in that country in late 1978. Although prices of petroleum sold by different petroleum exporting countries differ widely, the average price of petroleum imported into the United States has risen by 130 to 140 per cent since December 1978. In January 1981, President Reagan announced that prices of petroleum products would be decontrolled immediately. As a consequence, domestic petroleum prices in the United States have moved to world levels. Domestic natural gas prices will also be subject to gradual decontrol, moving to world levels as early as 1985 or, at the latest, 1987.

Given the sharp increase in the price of energy relative to the prices of other productive inputs, the prospects for productivity growth at the sectoral level are dismal. In the absence of any reduction in prices of capital and labor inputs during the 1980s, we can expect a decline in productivity growth for a wide range of U.S. industries, a decline in the growth of productivity for the U.S. economy as a whole, and a further slowdown in the rate of U.S. economic growth. To avoid a repetition of the unsatisfactory economic performances of the 1970s it is essential to undertake measures to reduce the price of capital input and labor inputs. The price of capital input can be reduced by cutting taxes on income from capital.[1] Similarly, payroll taxes can be cut in order to reduce the price of labor input. However, cuts in both taxes on capital and taxes on labor will require a reduction in overall government spending.

The prospects for changes in tax policy that would have a substantial positive impact on productivity growth in the early 1980s are not bright. Any attempt to balance the Federal budget during 1981 in the face of a slow recovery from the recession of 1980 will require tax increases rather than tax cuts. Higher inflation rates have resulted in an increase in the effective rate of taxation of capital. Payroll taxes are currently scheduled to rise in 1981. For these reasons it appears that a return to the rapid growth of the 1960s is out of the question. Even the moderate growth of the 1960s and early 1970s would be difficult to attain. In the absence of measures to cut taxes on capital and labor inputs, the performance of the U.S. economy during the 1980s could be worse than during the period from 1973 to the present.

For economists the role of productivity in economic growth presents a problem comparable in scientific interest and social importance to the problem of unemployment during the Great Depression of the 1930s. Conventional methods of economic analysis have been tried and have been found to be inadequate. Clearly, a new framework will be required for economic understanding. The findings we have outlined above contain some of the elements that will be required for the new framework for economic analysis as the U.S. economy enters the 1980s.

At first blush the finding that higher energy prices are an important determinant of the slowdown in U.S. economic growth seems paradoxical. In aggregative studies of sources of economic growth energy does not appear as an input, since energy is an intermediate good and flows of intermediate goods appear as both outputs and inputs of individual industrial sectors, canceling out for the economy as a whole.[2] It is necessary to disaggregate the sources of economic growth into components that can be identified with output and inputs at the sectoral level in order to define an appropriate role for energy.[3]

---

[1] An analysis of alternative proposals for cutting taxes on income from capital is presented by Auerbach & Jorgenson (1980).

[2] See, for example, Denison (1979).

[3] Kendrick (1961, 1973) has presented an analysis of productivity growth at the sectoral level. However, his measure of productivity growth is based on value added at the sectoral level, so that no role is provided for energy and materials inputs in productivity growth. For a more detailed discussion, see Jorgenson (1980).

Within a framework for analyzing economic growth that is disaggregated to the sectoral level it is not sufficient to provide a decomposition of the growth of sectoral output among the contributions of sectoral inputs and the growth of sectoral productivity.[1] It is necessary to explain the growth of sectoral productivity by means of an econometric model of productivity growth for each sector. Without such econometric models the growth of sectoral productivity is simply an unexplained residual between the growth of output and the contributions of capital, labor, energy, and materials inputs.

Finally, the parameters of an econometric model of production must be estimated from empirical data in order to determine the direction and significance of the influence of energy prices on productivity growth at the sectoral level.[2] From a conceptual point of view a model of production is consistent with positive, negative, or zero impacts of energy prices on sectoral productivity growth. From an empirical point of view the influence of higher energy prices is negative and highly significant. There is no way to substantiate this empirical finding without estimates of the unknown parameters of the econometric model of productivity growth.

The steps we have outlined—disaggregating the sources of economic growth down to the sectoral level, decomposing the rate of growth of sectoral output into sectoral productivity growth and the contributions of capital, labor, energy, and materials inputs, and modeling the rate of growth of productivity econometrically—have been taken only recently. Much additional research will be required to provide an exhaustive explanation of the slowdown of U.S. economic growth within the new framework and to derive the implications of the slowdown for future growth of the economy.

References

Auerbach, Alan J. & Jorgenson, Dale W.: Inflation-proof depreciation of assets. *Harvard Business Review 58* (5), 113–118, September–October 1980.

Berndt, Ernst R. & Jorgenson, Dale W.: Production structure. Chapter 3 in *U.S. energy resources and economic growth* (ed. Dale W. Jorgenson and Hendrik S. Houthakker), Energy Policy Project, Washington, 1973.

Berndt, Ernst R. & Wood, David O.: Technology, prices, and the derived demand for energy. *Review of Economics and Statistics 56* (3), 259–268, August 1975.

Berndt, Ernst R. & Wood, David O.: Engineering and econometric interpretations of energy-capital complementarity. *American Economic Review 69* (3), 342–354, September 1979.

Binswanger, Hans P.: The measurement of technical change biases with many factors of production. *American Economic Review 64* (5), 964–976, December, 1974.

---

[1] Gollop & Jorgenson (1980) have presented an analysis of productivity growth at the sectoral level based on the concept of output that includes both primary factors of production and intermediate inputs.

[2] Estimates of the parameters of an econometric model of sectoral productivity growth are presented by Jorgenson & Fraumeni (1981).

Binswanger, Hans P.: A microeconomic approach to induced innovation. *Economic Journal 84* (336), 940–958, December 1974.

Christensen, Laurits R. & Jorgenson, Dale W.: U.S. real product and real factor input, 1929–1967. *Review of Income and Wealth* Series 16 (1), pp. 19–50, March 1970.

Christensen, Laurits R., Jorgenson, Dale W. & Lau, Lawrence J.: Conjugate duality and the transcendental logarithmic production function. *Econometrica 39* (4), 255–256, July 1971.

Christensen, Laurits R., Jorgenson, Dale W. & Lau, Lawrence J.: Transcendental logarithmic production frontiers. *Review of Economics and Statistics 55* (1), 28–45, February 1973.

Denison, Edward F.: *Accounting for slower economic growth*. The Brookings Institution, Washington, 1979.

Diewert, W. Erwin: Aggregation problems in the measurement of capital. In *The measurement of capital* (ed. Dan Usher), pp. 433–538. University of Chicago Press, 1980.

Fraumeni, Barbara M. & Jorgenson, Dale W.: The role of capital in U.S. economic growth, 1948–1976. In *Capital, efficiency and growth* (ed. George M. von Furstenberg), pp. 9–250. Ballinger, Cambridge, 1980.

Gollop, Frank M. & Jorgenson Dale W.: U.S. productivity growth by industry, 1947–1973. In *New developments in productivity measurement and analysis* (ed. John W. Kendrick and Beatrice M. Vaccara), pp. 17–136. University of Chicago Press, Chicago, 1980.

Hicks, John R.: *The theory of wages*, Macmillan, London 1932. (2nd ed., 1963.)

Jorgenson, Dale W.: Accounting for Capital, in *Capital, Efficiency and Growth* (ed. George M. von Furstenberg), pp. 251–319, Ballinger, Cambridge, 1980.

Jorgenson, Dale W. & Fraumeni, Barbara M.: Substitution and technical change in production. In *The economics of substitution in production* (ed. Ernst R. Berndt and Barry Field), M.I.T. Press Press, Cambridge, forthcoming 1981.

Jorgenson, Dale W. & Griliches, Zvi: Issues in growth accounting: A reply to Edward F. Denison. *Survey of Current Business 52* (5), Part II, 65–94, 1972.

Jorgenson, Dale W. & Lau, Lawrence J.: *Transcendental logarithmic production functions*. North-Holland, Amsterdam, forthcoming 1981.

Kendrick, John W.: *Productivity trends in the United States*. Princeton University Press, Princeton, 1961.

Kendrick, John W.: *Postwar productivity trends in the United States, 1948–1969*. National Bureau of Economic Research, New York, 1973.

Samuelson, Paul A.: Prices of factors and goods in general equilibrium. *Review of Economic Studies 21* (1), 1–20, October 1953.

U.S. ENERGY PRICE DECONTROL: ENERGY, TRADE AND ECONOMIC EFFECTS

*Edward A. Hudson**

Dale W. Jorgenson Associates, Cambridge, Massachusetts, USA

Abstract

Energy price decontrol in the United States is analyzed in terms of its energy and economic effects. It is found that decontrol stimulates some increase in oil and gas supply together with a substantial reduction in demand. Decontrol affects real economic performance through several mechanisms; adjustments related to energy supply expansion and international trade tend to increase productivity while adjustments related to energy demand reduction lower productivity. Quantitative analysis of decontrol, using an energy-economy simulation model, shows that real economic growth is likely to be increased as a result of decontrol with the international trade effect being the dominant mechanism underlying these gains.

I. The Decontrol Issue

The prices of a wide range of goods and services, including petroleum, were brought under government control as part of President Nixon's economic policy in the early 1970s. This legal and administrative structure provided the basis for continuing to control petroleum prices in the face of the rise in world oil prices beginning in 1973–74. Government controls on the price of natural gas go back even further, ultimately to a series of court rulings in the 1950's. On these foundations, controls on petroleum and natural gas prices were well established by the mid-1970s. At that time, however, several major energy-related problems began to emerge. Natural gas shortages developed as consumption was spurred by low prices but resource development and production were retarded by low prices. Simultaneously, since petroleum price controls prevented the sharp increases in world oil prices from being reflected in domestic prices, there was little incentive to curtail petroleum use (in fact consumption and imports continued to rise) and there was restricted incentive to expand domestic supply activity. Arguments began to be put forward from several quarters about the desirability of energy price decontrol as a means of tackling these energy problems. In response, other groups made counter arguments that price decontrol would be ill-advised. A widespread and heated debate on energy price controls began.

* I wish to thank James E. Parrish for his excellent research assistance in performing the simulations used in this paper.

Typical arguments in favor of decontrol of energy prices fell into three categories—demand, supply and economic. First, rising prices to oil and gas consumers would reduce demand, thereby reducing imports (of petroleum) and shortages (of natural gas). This import reduction objective began to receive much attention in the aftermath of the 1973–74 world oil disruptions and even led to a "project independence" effort under President Ford. Second, higher prices to domestic oil and gas producers would stimulate exploration, reserve development and production, thereby enhancing the future supply of oil and gas and aiding in reducing import dependence and in easing shortages. Third, there were some arguments that decontrol would let markets operate more efficiently, leading to higher levels of production and real income. Typical arguments against decontrol fell into these same three categories. First, higher prices would result in very small demand reductions but at the cost of major new burdens on personal living standards. Second, higher prices to producers would mainly result in higher profits with only a very small increase in production. Third, the economic effects would be unfavorable, and there may be severe burdens placed on the living standards of the lower income groups. These arguments for and against decontrol can be reduced to a couple of central issues. Everyone agreed on the existence of energy benefits (demand reductions, supply expansions) although there was considerable disagreement about their magnitudes. There was more basic disagreement about the economic effects and whether any gains in real income would be generated by decontrol.

Energy price decontrol was enacted into legislation during the Carter administration. The price decontrol is phased—crude oil prices will gradually move towards market levels with full decontrol by October 1981; decontrol of natural gas prices is scheduled over a longer period, with wellhead price decontrol not fully attained until 1986. A "windfall profits tax" was introduced as part of subsequent legislation designed to transfer to the government part of the economic rent created by decontrol and also intended to accomodate political pressures against the possibility that oil companies and other energy producers might gain large profit increases from decontrol. These actions do not constitute complete price decontrol, in particular as many controls remain over the distribution and sale of petroleum products and natural gas, but they nonetheless represent a large measure of decontrol in the sense that crude oil prices are free to move to world levels.

II. Energy Effects of Decontrol

II.1. *Price Effects*

How has experience to date shed light on the actual outcome of decontrol? Examination of energy developments over the 1970s provides some information on the nature and magnitudes of the effects of price decontrol on the demand for and supply of petroleum and natural gas. Table 1 shows resource

Table 1. *U.S. energy prices*

	Refiner acquisition cost of crude oil[a]		Wellhead price of natural gas[b]	Retail prices of		Price of natural gas to households[e]
	Domestic	Imported		Gaso-line[c]	Heating oil[d]	
1974	7.18	12.52	30.4	0.46	NA	1.08
1975	8.39	13.93	44.5	0.45	0.30	1.21
1976	8.84	13.48	58.0	0.44	0.30	1.38
1977	9.55	14.53	79.0	0.44	0.33	1.60
1978	10.61	14.57	90.5	0.43	0.33	1.73
1979	14.27	21.67	114.4	0.53	0.40	1.95
1980	24.48	34.48	134.3	0.70	0.55	2.20

[a] Average price in dollars per barrel.
[b] Average price in cents per thousand cubic feet.
[c] Average price of gasoline in (1972) dollars per gallon. Prior to 1978, price is for leaded regular gasoline; 1978 and beyond price is for all gasoline.
[d] Average price for No. 2 heating oil in (1972) dollars per gallon.
[e] Average retail price of natural gas for residential heating in (1972) dollars per gallon.
[f] 1980 prices are for June for petroleum and for May for natural gas.

prices (crude petroleum and wellhead natural gas prices) and some demand or delivered prices for petroleum products and natural gas. The large disparity between domestic and imported crude oil prices is now being eliminated and prices to U.S. producers are approaching world levels. Similarly, average well-head prices for natural gas have begun to rise substantially. For producers, therefore, the move towards decontrol has resulted in significantly higher prices. Prices to oil and gas users are summarized by the real (i.e. constant dollar) retail prices for gasoline, heating oil and natural gas. Petroleum pro-duct prices remained virtually constant until 1978—price controls prevented the world energy price changes, the domestic natural gas shortages or anything else from making any impact on the relative price of petroleum products. Beginning in 1979, however, the move towards decontrol permitted a sub-stantial increase in relative prices; by mid-1980, real prices for petroleum pro-ducts were two-thirds above their 1978 levels. Real prices for natural gas began to rise in the mid 1970s as the controlled price was adjusted upwards and the initial moves towards decontrol have resulted in continuing increases. For consumers also the move towards decontrol has caused significant increases in the relative prices of petroleum products and natural gas.

II.2. *The Supply Response*

How have supply and demand responded to these price changes? The historical data suggests noticeable response on both sides of the energy system. Table 2 shows several indicators of U.S. oil and gas supply. Supply progresses through

Table 2. *U.S. supply and demand for oil and gas*

	Wells drilled[a]		Petroleum supply		Natural gas supply	
	Number[b]	Footage[c]	U.S.[d]	Imports[e]	U.S.[f]	Imports[g]
1970	28	139	20.4	6.9	21.7	0.8
1971	26	124	20.0	8.1	22.3	0.9
1972	27	135	20.0	9.8	22.2	1.0
1973	27	136	19.5	13.0	22.2	1.0
1974	32	151	18.6	12.7	21.2	0.9
1975	37	174	17.7	12.5	19.6	0.9
1976	40	182	17.3	15.2	19.5	0.9
1977	45	211	17.5	18.2	19.6	1.0
1978	47	227	18.4	17.1	19.5	0.9
1979	50	238	18.1	16.9	19.9	1.2
1980[h]	53	253	18.4	14.4	20.2	1.1

[a] Exploratory and development wells drilled in the U.S.
[b] Total wells drilled, in thousands.
[c] Total footage of wells drilled, millions of feet.
[d] U.S. production of crude oil, in quadrillion Btu.
[e] U.S. net imports of petroleum, in quadrillion Btu.
[f] U.S. production of natural gas (dry, marketed natural gas), in quadrillion Btu.
[g] U.S. net imports of natural gas, in quadrillion Btu.
[h] 1980 production is the January–June 1980 rate times two.

several stages—exploration to locate and prove new fields, development of known fields into sources of production, and actual production of oil and gas from developed fields. Producers have control over drilling activity and production rates, they do not have full control over finding rates. Therefore, only some aspects in the supply system are subject to producer control and can be expected to respond directly to the new incentive structure. Perhaps the principal decision variable is drilling. The table shows a clear trend in drilling. Drilling activity had been steadily declining from a peak in 1956 to a low point in 1971–73. Since 1974, the wellhead price of "new field" oil has been permitted to rise and this has been associated with a marked increase in drilling activity. Decontrol continues this upward movement in prices and has also been associated with large increases in drilling activity. In fact, drilling in 1980 is approximately double that of 1973 (despite a huge increase in drilling costs—costs per foot drilled rose by more than 150 % just between 1973 and 1980). This increased drilling has resulted in sufficient reserve additions and production capability to arrest the previous decline in the annual rate of oil and gas production. Annual production of petroleum was at its lowest level over the 1975–1977 period; production has since been stabilized at a slightly higher level. For natural gas, the period 1975–1978 saw low output levels; subsequently, production has been maintained at fractionally higher levels. In sum, rising oil and gas prices have caused a very large response by producers on the

Table 3. *U.S. demand for petroleum and natural gas*

	Petroleum use[a]	Consumption of[b]		Natural gas use[c]
		Gasoline	Fuel oil	
1973	34.8	6.7	3.1	22.5
1974	33.5	6.5	2.9	21.7
1975	32.7	6.7	2.9	19.9
1976	35.2	7.0	3.1	20.3
1977	37.2	7.2	3.4	19.9
1978	38.0	7.4	3.4	20.0
1979	37.1	7.0	3.3	20.5
1980[d]	34.9	6.6	2.9	22.2

[a] Primary input of petroleum, in quadrillion Btu.
[b] Consumption of gasoline and of distillate fuel oil, in million barrels a day.
[c] Primary input of natural gas, in quadrillion Btu.
[d] 1980 use is estimated as the actual January–June level times two.

main variable under their control—drilling rates. The effect of this on annual production has been large enough, even at this relatively early stage, to halt the previous decline in annual production levels of petroleum and of natural gas.

II.3. *The Demand Response*

There has also been a response in energy use, particularly for petroleum, to higher relative prices. Table **3** shows primary input of petroleum and of natural gas and also consumption rates of gasoline and fuel oil. Natural gas use has not declined in response to recent price rises; in fact, natural gas use has followed the movements in supply. The situation is that natural gas use, particularly in the mid-1970s, has been supply constrained. Prices do not align supply and demand, rather allocation schemes by gas utilities serve to restrain demand to the quantity of gas available. Since demand is not observable, the price impact on demand cannot be directly seen. In petroleum, however, supply has been able to adjust to demand at prevailing prices (the import quantity adjusts to accomodate and demand not satisfied by U.S. production) so the impact of price on demand can be observed directly. Petroleum consumption increased to a peak in 1978 but, since then, has dropped sharply. Gasoline use follows very closely the pattern of price changes—demand rose until 1978 (a period of unchanged relative prices) but then, as real prices started to rise, fell substantially (1980 consumption is 11 % below the 1978 level). Distillate fuel oil consumption has also fallen, reversing the rising trend up to and including 1978. Again, the turning point in use patterns coincided with the beginning of the increase in the relative price of fuel oil. These declines in gasoline and fuel oil use have occurred despite continued growth in production, real incomes, the number of cars and the number of houses. All these aspects of petroleum consumption are consistent with a significant and rapid response to higher relative prices of petroleum products.

II.4. *Petroleum Imports*

Petroleum demand has decreased in recent years while U.S. supply has stopped decreasing, and even shown a slight rise. Since imports make up the balance between demand and domestic supply, the changes in these variables have caused a large change in imports. Petroleum imports peaked in 1977 at 18.2 quadrillion Btu or approximately 9 million barrels a day or over 50 % of total consumption. Since then, particularly as a result of the reductions in total petroleum use, imports have fallen sharply. In 1980, imports are estimated to be around 7 million barrels a day, or less than 45 % of total demand. Thus, while the U.S. continues to obtain a substantial part of its petroleum from imports, the degree of import dependence has been markedly reduced over the period of energy price decontrol and domestic price increases. The demand-supply changes have had different implications for natural gas. Here, any imbalance between demand and domestic supply at prevailing prices has been taken up by forced, i.e. non-price, demand reductions and not by imports (imports have remained almost constant at a 1 quadrillion Btu annual level). The effects of domestic supply increases and any demand reductions that might have occurred as a result of the price rises have been felt in a reduction of the shortage and in an increase in the level of demand that is able to be satisfied.

II.5. *Overview of Energy Effects*

These energy developments have been generally in line with the predictions of the proponents of price decontrol. Particularly for petroleum, there has been a large reduction in levels of petroleum use, despite increasing population and economic activity. In short, there has been a significant, and rapid, demand response to higher petroleum prices. There also has been a noticeable supply response in both oil and gas. The essence of the response has occurred in the main variable under the control of producers—drilling. Exploratory and development drilling rose markedly in response to higher oil and gas prices. The production effects have been sufficient to halt the previous decline in domestic production. Finally, demand reduction and supply expansion have had a major effect on the difference between demand and domestic supply—petroleum imports have been substantially reduced and the unsatisfied demand for natural gas eased.

III. Economic effects of Decontrol: Qualitative Analysis

III.1. *Introduction*

This historical review is useful in gaining some insight into the energy effects of changing energy prices since changes from past trends were so distinct, in both cause and effect variables, that interdependence can be observed. The analysis of the economic effects cannot be so direct, however, as there is no basis for attributing recent economic developments solely to the energy price changes. What is needed is a comparison of what economic performance would be with decontrol and without decontrol, but with all other exogenous influences being held unchanged. Such information can be provided by simulation analysis using empirically based models of the energy and economic systems. This approach is used to develop quantitative estimates of the effects of price decontrol. But, as a prelude to this, conventional arguments about the economic effects of decontrol are appraised and the mechanisms through which decontrol affects the economy are analyzed in a qualitative fashion.

The conventional energy price decontrol-macroeconomic linkage argument is that decontrol will lead to higher real GNP. The arguments or mechanisms involved appear to be that decontrol will mean improved energy market efficiency and that a reduced level of market imperfections will result in improved productivity in the economy as a whole. If this is indeed the core of the standard argument, it is hardly conclusive. It does not deal with how consumers can be better off in the face of higher prices, nor does it deal with how producers will create more output once they cut back on energy use. Further, removal of some, among many, market imperfections does not necessarily lead to a preferred situation, as demonstrated by the theory of the second best. It must be concluded, therefore, that the conventional argument does not establish, one way or the other, any effect of energy price decontrol on real GNP.

III.2. *Economic Productivity and Energy Demand*

In fact, careful examination shows that several types of linkages between energy price decontrol and macroeconomic potential, as indicated by real GNP, do exist and can be analyzed. It emerges that some mechanisms act to increase real GNP, some to decrease it. The three key mechanisms are— productivity effects of energy use changes, productivity effects of energy supply changes, and effects of international trade adjustments. In addition, there are dynamic effects, in particular those operating through investment and the growth of productive capacity, that can affect future growth of real incomes and output.

Higher energy prices, arising from price decontrol, lead to a pervasive set of adjustments by energy purchasers. Spending patterns by consumers are adjusted away from now relatively more expensive energy and energy-intensive goods as consumers seek to gain maximum preference levels, subject to their strictly limited incomes. The relative shifts from energy and from large automobiles towards services are important examples of the restructuring of purchase patterns induced by rising energy prices. Similarly, producers redirect their input patterns to economize on energy and energy-intensive goods as they seek to minimize average production costs in the face of the new price structure. The changing of input patterns away from energy and towards capital and labor are central examples of the substitution responses in production activities. In sum, there is a change in what is produced (final demand) and how it is produced (input patterns), in both cases the change being away from energy. Other inputs must increase in relative importance in place of energy and the principal overall adjustment is for an increase in capital and labor input per unit of output. This is equivalent to a reduction in the average output per unit of capital and labor input or, in practical terms, a reduction in the average level of capital and labor productivity throughout the economy. It is possible, although not certain, that these restructuring processes, within industries and the reallocation of resources between industries, will in themselves involve wastage and inefficiencies; such effects would compound the productivity impact. Also, productivity growth is likely to be hurt by the intersectoral shifts in production. In particular, the shift of spending and production towards services involves increasing the relative importance of a sector with comparatively low levels of and slow growth of productivity. All of these effects work in the same direction: the change of demand patterns in response to higher energy prices involves a productivity cost; future productivity levels and growth rates are lower than they would otherwise have been.

III.3. *Productivity and Energy Supply*

Higher prices to domestic producers of petroleum and natural gas induce an increase in exploration and development activity and ultimately in the production of these fuels. Each barrel of expanded domestic production means one barrel less of imported oil. If the marginal cost of this expanded domestic production is less than the world oil price then this supply substitution will provide a given oil supply at a lower resource cost than in the controlled price situation. If domestic production is subject to increasing cost (an upward sloping marginal cost curve) and if production is pushed to the point where marginal cost equals price then price decontrol, with the associated rise in prices to the world level, will induce an increase in domestic output at a smaller cost than that of the imports displaced. This release of economic resouces from energy supply leaves more inputs to the rest of the economy, permitting an increase in total output, i.e. permitting a rise in overall productivity. However, the magnitude of this effect depends on the size of the supply elasticity. The historical data reviewed above suggests that the supply elasticity is rather less than the demand elasticity; in such conditions it may well be the case that this supply-related productivity improvement is outweighed by the demand-related productivity reduction.

III.4. *International Trade Effects*

The international value of the dollar and U.S. international trade will certainly be affected by domestic energy price decontrol; these changes will in turn

Table 4. *U.S. international trade, 1970–1980*

	Prices of[a]		Quantities of[b]		Balance of trade[c]
	Imports	Exports	Imports	Exports	
1970	0.89	0.93	87	92	3.9
1971	0.94	0.97	90	93	1.6
1972	1.00	1.00	100	100	− 3.3
1973	1.18	1.16	105	120	7.1
1974	1.71	1.48	102	128	6.0
1975	1.88	1.64	89	124	20.4
1976	1.93	1.70	106	132	8.0
1977	2.11	1.79	116	135	− 9.9
1978	2.22	1.90	129	150	− 10.3
1979	2.56	2.15	135	165	− 4.6
1980[d]	3.09	2.39	132	177	− 2.5

[a] Price indices, on 1972 = 1.0, of U.S. imports and exports, U.S. National Income and Product Accounts definitions.
[b] Trade volumes, in constant 1972 dollars expressed relative to 1972 = 100; U.S. National Income and Product Accounts definitions.
[c] The balance of trade is net exports, in National Income and Product Accounts terms, in billions of dollars.
[d] 1980 figures are for the second quarter at an annual rate.

impact real incomes within the U.S. Higher domestic energy prices will cause a reduction in demand, some increase in domestic production and, therefore, a decline in the quantity of energy imported. Since import prices remain unchanged, this translates directly into a lower level of import payments and a higher trade surplus (or lower deficit). These changes will affect the international value of the U.S. dollar, leading to a revalued dollar. This, in turn, is likely to lead to a reduction in U.S. export volumes, as U.S. exports are placed at a relative price disadvantage and as exporting becomes less profitable to U.S. producers. But, the goods and services previously exported, or at least the inputs to these goods and services, now become available to the domestic market. The volume of domestic consumption, investment and government purchases can increase, even within existing constraints on the availability of economic inputs. In short, the reduction in imports and in export volumes permits a rise in the other components of real GNP, the overall productivity of the economy rises and the level of achievable real GNP rises. This is a potentially powerful mechanism whereby decontrol of domestic energy prices will feed back to raise real incomes and real GNP.

The strength of this international trade effect of decontrol depends greatly on the response of export quantities to changes in the trade balance. Recent historical data can be reviewed to throw some light on the behavior of export quantities. Table 4 shows U.S. import and export price and quantity indicators for the 1970–1980 period. Import prices showed large rises in 1974 and in 1979–80 (both of which were associated with large increases in world oil prices). Subsequent adjustment in the value of the dollar caused rises in export prices over and above general inflation. Import quantities have shown a response to the price rises, for example imports have been level since 1978. Export quantities, in contrast, have shown a large increase. While this evidence is not conclusive, as other forces simultaneously involved were not explicitly allowed for, it is certainly consistent with a sizeable price-induced response in export quantities. Also, the trade balance data in the table shows that the trade balance has stayed within very narrow limits, despite the large rises in import prices and import payments. This implies that trade quantities have adjusted sufficiently strongly and sufficiently rapidly to keep the trade balance within narrow bounds. This whole set of evidence is consistent with a quantitatively significant mechanism linking import changes, via trade balance and dollar value changes, to export quantity changes.

III.5. *Overview of Economic Effects*

Three important, but distinct, types of macroeconomic feedbacks from energy price decontrol have been identified:

(1) Domestic demand reduction; this is associated with productivity and real GNP losses;

(2) domestic supply expansion; this is associated with productivity gains;

(3) import reduction and trade effects; these are associated with productivity and real GNP gains.

These mechanisms involve one set of pressures towards real income losses and two towards real income gains. A qualitative analysis cannot tell which pressures will dominate and whether the overall impact will be one of real GNP gains or of losses, it can only establish that either direction is a possibility. The next step, therefore, is to perform quantitative analyses to permit these different pressures to be considered simultaneously and to permit the overall direction and magnitude of impact to be ascertained.

IV. Quantitative Analysis of Economic Effects

IV.1. *The Simulations*

The Hudson–Jorgenson energy-economy model system was used to simulate two futures, one with continued energy price controls, the other with decontrol, over the period 1980 to 1990. (The structure and nature of this model system are outlined in the Appendix.) The controls case involved the retention of the structure of price controls and associated allocation schemes that emerged during the 1970's; the decontrol case involved phased price decontrol and the windfall profits tax, plus associated tax changes, corresponding to legislation

Table 5. *Decontrol impact on prices and supply, 1980 to 1990*

	Continued controls	Decontrol	Difference (%)
Petroleum prices[a]			
1985			
Crude oil	24.60	31.96	30
Gasoline	35.75	44.05	23
Distillate	26.76	33.61	26
1990			
Crude oil	29.31	36.40	24
Gasoline	40.93	49.09	20
Distillate	32.12	38.49	20
U.S. oil and gas production[b]			
1985			
Crude oil	17.6	18.7	6
Natural gas	17.9	18.2	1
1990			
Crude oil	16.9	19.6	16
Natural gas	18.4	18.7	2

[a] Crude oil price is the average refiner acquisition price; gasoline and distillate fuel oil prices are average prices at the city gate; prices are in $ (1979) per barrel.
[b] U.S. production of crude oil and natural gas in quadrillion Btu.

passed in the 1978–1980 period. In each case, the world crude oil price was assumed to rise from (in constant 1979 dollars per barrel) 30 in 1980 to 37 in 1990. This oil price assumption and detailed analyses of petroleum and natural gas price developments were based on recent work done by the Energy Information Administration (EIA) of the U.S. Department of Energy. The calculations about product prices, in both controlled and decontrolled situations, incorporate a wealth of detail on economic, regulatory, legislative and institutional features. For example, price calculations in the controlled case take account of all the many different legal categories of crude oil production, each category having a different price. Table 5 indicates the impact of decontrol on petroleum prices. Decontrol causes the average price of crude oil to rise by 30 % in 1985 and by 24 % in 1990, relative to the continued controls case. The domestic supply response estimated by EIA is also summarized in this table. The higher prices, even with the windfall profits tax, induce additional exploration and development activity, resulting in higher production of both oil and natural gas. The estimated gas production response is small (only 1 or 2 %) but the oil production response is larger and increasing over time, with output in 1990 under decontrol estimated at about one-sixth higher than under continued controls.

Table 6. *Composition of real final demand*

	Continued controls	Decontrol	Difference (%)
1985			
Agriculture, construction	0.1191	0.1196	0.4
Manufacturing	0.3842	0.3829	− 0.3
Transportation	0.0410	0.0402	− 2.0
Services, trade	0.4108	0.4138	0.7
Energy	0.0450	0.0436	− 3.1
Total	1.0000	1.0000	
1990			
Agriculture, construction	0.1158	0.1164	0.5
Manufacturing	0.3936	0.3907	− 0.7
Transportation	0.0436	0.0422	− 3.2
Services, trade	0.4049	0.4101	1.3
Energy	0.0421	0.0405	− 3.8
Total	1.0000	1.0000	

IV.2. *Prices and Final Demand*

The initial impact of the rise in energy prices is on purchase patterns, both those by consumers and those by producers. Energy use is the most directly affected as purchase and input patterns are shifted away from energy and energy-intensive goods and services. Demand for petroleum products is 5 % less in 1985 and 6 % less in 1990 than under continued controls. Natural gas use is reduced by about 1 % in each year. But other purchases are also affected both because spending shifts away from now relatively expensive energy-intensive products and because spending is shifted towards low energy-intensive goods and services. Final demand (of which the largest component is personal consumption but which also includes investment, government purchases and exports) is restructured. Table 6 shows the broad composition of final demand, and how this composition changes, for 1985 and 1990. Energy (petroleum products, natural gas and electricity) becomes relatively less important in the total quantity of final purchases, being between 3 and 4 % less than in continued control conditions. Energy-intensive products also decline in importance. Commercial transportation is reduced by up to 3 % and manufactured goods are also slightly reduced. In the place of these reductions, agriculture, construction and services purchases are projected to rise. The largest increases, in both percentage terms and in absolute size, is in services.

IV.2. *Production Inputs and Outputs*

Producers adjust input patterns, moving to reduce the relative importance of the now more expensive energy and energy-intensive inputs. Table 7 summa-

Table 7. *Structure of inputs to non-energy production*

	Continued controls	Decontrol	Difference (%)
1985			
Capital	0.1305	0.1308	0.4
Labor	0.2994	0.3002	0.3
Energy	0.0353	0.0347	−1.7
Materials	0.5349	0.5344	−0.1
Total	1.0000	1.0000	
1990			
Capital	0.1280	0.1280	0.8
Labor	0.2836	0.2840	0.1
Energy	0.0357	0.0350	−2.0
Materials	0.5527	0.5520	−0.1
Total	1.0000	1.0000	

rizes aggregate input patterns for non-energy production throughout the economy. The energy intensity of production falls (on average by about 2 %) and there is a very slight decline in the relative role of intermediate materials. Both capital and labor become relatively more important as productive inputs. Input patterns within each industry are altered, typically towards capital and labor. In addition, there are relative shifts between industries and this impacts the capital and labor intensity of production as a whole. Initially, the greater response is in labor input, in part because labor is more flexible than capital, but over time capital input can adjust to these pressures. In all years, however, capital intensity and labor intensity of production are increased as part of the adjustment to higher energy prices.

The change in final demand patterns (what is produced) and in input patterns generate a further change in the mix of gross output in the economy. Thus, the relative growth rates and relative sizes of different industries are altered. Production and employment patterns are correspondingly altered. Table 8 shows these industrial shifts for the five broad sectors. The directions of these changes are similar to those in final demand although the magnitudes are slightly less. Energy, transportation and manufacturing have reduced growth rates while agriculture, construction and services increase at faster rates. In terms of the absolute adjustments, the greatest single change is the increase in output from and employment in the service industries.

IV.4. *Trade Effects*

Decontrol also has large international trade effects. The decline in U.S. petroleum demand coupled with the rise in domestic production means that import requirements are substantially reduced. In 1985, imports are reduced

Table 8. *Composition of total real demand*

	Continued controls	Decontrol	Difference (%)
1985			
Agriculture, construction	0.0922	0.0925	0.3
Manufacturing	0.4517	0.4506	− 0.2
Transportation	0.0517	0.0512	− 1.0
Services, trade	0.3438	0.3454	0.5
Energy	0.0606	0.0602	− 0.7
Total	1.0000	1.0000	
1990			
Agriculture, construction	0.0907	0.0911	0.4
Manufacturing	0.4617	0.4598	− 0.4
Transportation	0.0541	0.0533	− 1.5
Services, trade	0.3383	0.3411	0.8
Energy	0.0553	0.0547	− 1.1
Total	1.0000	1.0000	

from 15.2 quadrillion Btu with continued controls to 12.1 quads with decontrol, a 20 % reduction. In 1990 the decline is 30 %, from 16.6 to 11.7 quadrillion Btu. Since the prices for oil imports are not affected by decontrol, the quantity reduction causes a large drop in import payments—a reduction of about \$35 bn in 1985 and \$95 bn in 1990. As oil import payments represent over one-third of the value of U.S. imports of goods and services, these changes constitute a relatively large reduction in total U.S. import payments. The improved trade balance will lead to adjustments in the value of the dollar. Exports will then be affected; in fact they will be reduced as a result of decontrol. These trade adjustments were modeled on the basis of a trade constraint—the current dollar trade balance was required to be the same under decontrol as under continued controls. As a result of this constraint and the lower import payments, export quantity growth is slowed, although it is nonetheless positive. By 1990, real exports of goods and services are estimated to be 11 % less under decontrol. This export change affects industries differently and contributes to the restructuring discussed above. Also, these trade changes contribute to the macroeconomic effects analyzed below.

IV. 5. *Macroeconomic Impacts*

These structural changes permit the energy intensity of economic production to be reduced but these changes themselves impact overall economic performance, as indicated, for example, by real GNP. The mechanisms underlying these macroeconomic effects have been outlined above. Energy supply expansion, energy use reduction, sectoral shifts, increases in capital and labor

Table 9. *Effects of decontrol on real GNP and its components*

Figures in billions of 1972 dollars

	Continued controls	Decontrol	Difference (%)
1985			
Consumption	1 065.1	1 078.2	1.2
Investment	244.0	250.8	2.8
Government	300.9	300.9	—
Exports	153.5	142.7	−7.0
Imports	120.8	118.1	−2.2
Net exports	32.6	24.6	
GNP	1 642.7	1 654.5	0.7
Real disposable personal income per capita	4 945	5 035	1.8
1990			
Consumption	1 225.9	1 254.7	2.3
Investment	276.0	289.1	4.7
Government	345.5	345.5	—
Exports	204.8	182.0	−11.1
Imports	159.3	155.0	−2.7
Net exports	45.1	27.0	
GNP	1 892.5	1 916.3	1.3
Real disposable personal income per capita	5 484	5 652	3.1

intensity of production, import and export changes are all important elements of the adjustment process as they affect real GNP. The macroeconomic impacts are summarized in Table 9. Real GNP is projected to increase as a result of decontrol with the increase relative to the continued controls case being permanent and rising over time. Real GNP in 1985 is projected to be $(1972)12 bn or 0.7 % higher as a result of decontrol; by 1990, the difference is $(1972)24 bn or 1.3 %. The increase in use of final product is concentrated in consumption and investment. Government purchases remain fixed by assumption, so they don't share the rise in output. Real net exports (i.e. exports in 1972 dollars) decline since the export volume reduction exceeds the import volume reduction corresponding to equal current dollar trade changes (as import prices have risen more than export prices since 1972). The rise in real consumption is associated with higher real disposable personal incomes—up 1.8 % in 1985 and 3.1 % in 1990. Personal savings increase (the income rise exceeds the consumption rise). This is a permissive factor behind the substantial rise in real investment. In turn, higher investment means a more rapid rate of capital accumulation and further increases in future productive capacity, one reason for the permanently higher real GNP path associated with decontrol.

V. Productivity and Trade Effects

V.1. *Specification of the Simulations*

The previous qualitative analysis identified several key mechanisms whereby the adjustments brought about by energy price decontrol would affect real GNP and its growth. These mechanisms involved expanded domestic energy supply, reduced energy use and international trade adjustments. (In turn, investment and capital effects would compound initial changes.) It appeared that the productivity reducing effects of reduced energy use would dominate the improvements in energy supply productivity, leaving an adverse impact on overall economic productivity. Thus, the macroeconomic effects of decontrol might be grouped into two: internal productivity reductions, stemming from these energy use and supply changes, and productivity improvements resulting from the reallocation of resources away from exporting towards producing for domestic consumption and investment. These two effects, which might be labeled the productivity and trade effects, operate in opposite directions. The quantitative analysis just presented established that real GNP increased as a result of decontrol, implying that the trade effect outweighted the internal productivity effect. However, it is possible to perform additional simulations that explicitly separate out the productivity from the trade effects.

The decontrol simulation was performed with no trade constraint at all, but with all other conditions on energy prices and supply as in the previous decontrol case. This specification is equivalent to giving away, without benefit, all the imported petroleum released as a result of higher domestic supply and lower demand. Analytically, this specification involves no international trade effects, i.e. it limits the effects of decontrol to the internal productivity effects. Comparison of this new projection to the continued controls case shows the macroeconomic impacts of the internal productivity effects of decontrol; comparison of this new case with the previous decontrol case shows the impacts of the international trade effects of decontrol.

V.2. *The Separate Effects*

The results of the simulations to separate the productivity and the trade effects are presented in Table 10. For each year, the same features are apparent:

(*a*) The effect of higher energy prices on internal productivity is negative, i.e. in the absence of any trade adjustment, energy price decontrol causes a reduction in real GNP;

(*b*) The effect of introducing the trade constraint, i.e. of receiving economic benefit from the reduction in oil imports, is positive, leading to an increase in real GNP;

(*c*) In quantitative terms, the trade effect is far larger than the productivity effect.

Table 10. *Productivity and trade effects of decontrol*

Figures in billions of 1972 dollars

	1985	1990
A. Continued controls	1 642.7	1 892.5
B. Decontrol, no trade constraint	1 640.8	1 885.0
C. Decontrol	1 654.5	1 916.3
Separate effects		
Productivity (B − A)	− 1.9	− 7.5
Trade (C − B)	13.7	31.3
Total (C − A)	11.8	23.8

These findings are certainly consistent with the previous qualitative results. But, they permit quantitative magnitudes to be placed on the effects so that, in 1990 for example, the $(1972)31 bn trade effect is sufficient to offset the $(1972)8 bn loss from the internal productivity effect and to leave a $(1972)24 bn net gain from decontrol.

V.3. *Impact of Partial Trade Adjustment*

What would be the net outcome of decontrol if there were only partial adjustment in foreign trade so that only some of the reduction in oil imports was reflected in reduced exports? Since partial trade adjustment is equivalent to giving away, without benefit, part of the decontrol-induced oil savings it is clear that the macroeconomic gain from decontrol in this case will be less than that from decontrol in the general case with full trade adjustment. A series of simulations were performed to address this issue in a quantitative manner. Table 11 shows the results for four different decontrol situations—decontrol with 0, 20, 40 and 100% trade adjustment (where 0% adjustment is for no trade change other than lower imports and 100% adjustment is for sufficient trade change that the current dollar trade balance is restored to its level in the continued controls case). These results show that the size of the real GNP gain from decontrol is directly related to the degree of trade adjustment. In 1990 for example, decontrol with no trade constraint has a real GNP effect of − 7.5 (in $(1972)bn), the productivity effect alone, while the effect is − 1.8 for 20% trade adjustment, becoming positive to 4.3 for 40% trade adjustment and rising to 23.8 for full adjustment. These findings show that for 20% or less trade adjustment the productivity effect of decontrol dominates the macroeconomic outcome and results in an adverse overall effect of decontrol. But, for anything more than about 20% trade adjustment, the trade effect is the dominant force and the previous conclusions about the beneficial real GNP impact of decontrol hold true. The previous review of historical trade adjustments suggested that considerably more than 20% adjustment could be ex-

Table 11. *Real GNP effects of decontrol with different trade constraints $(1972) bn*

	1985	1990
Continued controls	1 642.7	1 892.5
Decontrol with trade adjustment of		
0 %	1 640.8	1 885.0
20 %	1 644.1	1 890.7
40 %	1 647.2	1 896.8
100 %	1 647.2	1 896.8
100 %	1 654.5	1 916.3
Effect of trade constraint for adjustment of		
0 %	0	0
20 %	3.3	5.7
40 %	6.4	11.8
100 %	13.7	31.3
Effect of decontrol for trade adjustment of		
0 %	− 1.9	− 7.5
20 %	1.4	− 1.8
40 %	4.5	4.3
100 %	11.8	23.8

pected in practice. Therefore, it might generally be expected that the trade effect would dominate and that energy price decontrol would result in a relative increase in real GNP.

VI. Summary and Conclusions

Energy price decontrol has been examined in terms of its historical effects, i.e. those observed to date, and of its effefts projected for the future. The energy effects are not surprising—demand has been reduced and domestic production increased—although it is striking that these effects appear to be both substantial in size and rapid in time. The crux of the analysis, however, has involved the macroeconomic effects of decontrol. Three key mechanisms were identified whereby price decontrol leads to adjustments in real GNP: (a) reduced energy use involves a productivity cost and lower real GNP, (b) expanded domestic energy production displaces imports and has a productivity benefit, tending to raise real GNP, (c) reduced imports are associated with slower export growth, releasing resources to production for domestic use and tending to raise real GNP. These linkages mean that decontrol sets in motion some forces working to raise real GNP and others to lower it. The net effect was analyzed using an energy-economy simulation model. This approach showed first, that the internal productivity effects of decontrol (i.e. (*a*) and (*b*) together) were to reduce the level and growth of real GNP but that this reduction was far outweighed by the overall productivity improvement associated with the trade effect, (*c*). Thus, decontrol has a favorable macroeconomic impact with the key being the induced adjustments through international trade. Without these trade adjustments, decontrol has an unfavorable effect although it was shown that the effect would be favorable for anything more than 20 % adjustment of export revenue to the change in import payment. (Also, this result does not mean that rising world oil prices have beneficial effects since, unlike decontrol, higher world oil prices can involve additional import payments.)

To summarize, the central features of this study are two. First, the issue of energy price decontrol was analyzed on both energy grounds and macroeconomic grounds; it was concluded that decontrol has a favorable energy impact (lower demand and increased domestic supply), and that decontrol was likely to result in higher levels of and faster growth of real GNP. Second, the mechanisms in the energy-economy linkage were clarified and were systematically analyzed in terms of nature and quantitative importance. In particular, a new mechanism—the international trade effect and feedback—was revealed and determined to be of central importance in the economic impacts of energy price changes.

Appendix

Summary of the Hudson–Jorgenson Energy-Economy Model

The Hudson–Jorgenson system is a structural model of the U.S. economy and economic growth. For each year, it analyzes economic activity on a sectoral basis then integrates these sectors into a consistent whole. There are several producing sectors of which four are non-energy and the rest cover energy extraction and processing. These sectors are agriculture, non-fuel mining and construction; manufacturing; transportation; services, trade and communications; coal mining; crude petroleum and natural gas; petroleum refining; electric utilities; and gas utilities. In addition, new technology energy supply and conversion activities are explicitly included under five categories: coal substitutes, liquid fuel substitutes, electricity substitutes, and others. New sources of supply are specified in further detail then aggregated into these five categories. Inputs into production are provided by the nine main producing sectors and by three types of primary factors—capital, labor and imports. And there are four categories of final demand for goods and services: personal consumption expenditures, investment, government purchases and exports.

These activities are organized into a matrix of interindustry transactions with 12 supplying sectors and 18 purchasing sectors. Within this interindustry framework, balance or consistency is required to hold in several senses. First, price formation must be such that sectoral output prices cover average costs of production, including a normal rate of return. Second, quantities must be such that, for every sector, the output of a sector exactly matches the quantity of that good or service required for input into other producing sectors together with the quantity used to satisfy final demand. Third, prices and quantities must be such that the revenue received by a sector is exactly accounted for by payments to inputs, including income to capital, and by payments to governments. Fourth, the demands for capital and labor inputs must be consistent with the supplies of these resources and also imports are constrained by the available revenue from exports together with a limited foreign deficit.

Activity patterns within this framework are represented by econometric models. The sub-models for consumption and for each of the non-energy producing sectors incorporate the patterns of behavioral and technical responses observed for these activities. This approach gives a flexible yet realistic and consistent representation of economic behavior. These sub-models provide a framework for the analysis of output price formation and, with the above consistency conditions, determine the system of relative prices characterizing the economy. Also, these sub-models determine that pattern of input purchases, the input–output coefficients, that, of feasible input patterns, represent the cost minimizing pattern given prevailing prices. I.e., the input–output coefficients are endogenous, being functions of, inter alia, relative prices. Similarly, the pattern of consumption spending is modeled in a flexible manner with consumer spending on each type of good or service being dependent on, inter alia, income and relative prices. These features provide for the incorporation of economic behavior, substitution or complementarity relationships between inputs, and the nature of adjustment in the pattern of consumer expenditure.

As a result, some of the principal adjustment mechanisms in the economic system are explicitly incorporated in the model. For energy, these features incorporate the analysis of the patterns of use of energy as an input to each type of production as well as the patterns of energy use in consumption and other final demand activities. For the economy in general, this specification provides for the incorporation of a full supply representation of production, including input requirements, input constraints, input adjustment possibilities and productivities. Further, the general equilibrium specification ensures that the direct and indirect effects throughout the economy of energy (or other) changes are allowed for.

This system is a dynamic equilibrium model of the U.S. economy. For each of the commodities endogenous to the model, the model incorporates an algorithm of price formation and a balance between demand and supply that determines relative prices. The pattern of economic activity in each year is consistent with these prices, with activity patterns based on observed patterns of substitution and other responses by producers and consumers. In addition, the model includes a balance between saving and investment that determines the rate of return and the rate of growth of capital stock. Economic growth is modeled as a sequence of one-period equilibria determining demand and supply and relative prices for all commodities. Investment in each period determines the level of capital stock available in the following period. The principal dynamic adjustment to energy changes is analyzed by tracing through the impact on future levels of capital stock.

References

Economic Report of the President, U.S. Government Printing Office, Washington D.C., 1980.

Energy Information Administration, U.S. Department of Energy, *Annual Report to Congress 1979, Volume Two: Data*, Washington D.C., 1980.

Energy Information Administration, U.S. Department of Energy, *Annual Report to Congress 1979, Volume Three: Projections*, Washington D.C. 1980.

Energy Information Administration, U.S. Department of Energy, *Monthly Energy Review*, Washington D.C., September 1980.

Hudson, E. A. & Jorgenson, D. W.: The economic impact of policies to reduce U.S. energy growth. *Resources and Energy 1*, 205–229, 1978.

Hudson, E. A. & Jorgenson, D. W.: Energy policy and U.S. economic growth. *The American Economic Review 68* (2), 118–123, May 1978.

Hudson, E. A. & Jorgenson, D. W.: *The long term interindustry transactions model: A simulation model for energy and economic analysis*. Federal Preparedness Agency, General Services Administration, Washington D.C., 1979.

U.S. Department of Commerce, *Survey of Current Business*, Washington D.C., August 1980.

THE MACROECONOMIC EFFECTS OF THE 1979/80 OIL PRICE RISE ON FOUR NORDIC ECONOMIES

*Ian Lienert**

OECD Secretariat, Paris, France

Abstract

This article reviews the main short-term macroeconomic consequences following a rise in the price of OPEC oil. Output is postulated to be reduced to the extent that the recipients of increased oil revenues (either OPEC or domestic oil producers) increase savings, and to the extent to which domestic real income is redistributed. Using the INTERLINK model of the world economy, this proposition is tested for Denmark, Finland, Norway and Sweden. Under the assumptions of unchanged exchange rates and policies (including no domestic respending of Norway's incremental revenues), the hypothetical results suggest that consequent to the 1979/80 oil price rises, Finland's output loss is much smaller than the other three OPEC-dependent countries because its terms of trade loss is largely offset by increased exports to the USSR.

I. Introduction

Several authors have discussed the world-zone macroeconomic implications of a rise in OPEC oil prices; see Corden (1976), Fried & Schulze (1975), OECD (July, 1980) and Okun (1975). Others, such as Eckstein (1978) and Mork & Hall (1978), have used various quantitative models to simulate the 1974/75 and 1978/79 oil price shocks, but these have generally been applied to large economies. For small, more open economies dependent on imported oil, there seem to be fewer studies; see Economic Council of Denmark (1979), Johnson & Klein (1974) and Norwegian Ministry of Finance (1980). A major reason is that for such countries, a domestic model with a sufficiently detailed energy sector needs to be embedded in a world model which incorporates both oil and non-oil trade, in order to capture output-reducing effects originating directly from oil producers with low spending propensities and indirectly from reduced demand of other industrialised and developing countries.

* The author has benefited from helpful comments from colleagues of the International Energy Agency and the OECD Secretariat, especially Messrs. Dean, Larsen, Llewellyn, Peura (now Bank of Finland) and Samuelson, together with other economists in Scandinavian capitals. Remaining errors and shortcomings are entirely the responsibility of the author. Any views expressed are the author's and do not necessarily reflect opinions of the OECD.

In Part II of this study, the determinants of macroeconomic changes following an oil-induced shift in the distribution of world income are reviewed. It is argued that, after an OPEC oil price rise, a deflationary force on output would arise not only from a redistribution of real income to OPEC, who typically do not immediately spend their additional income, but also from a domestic real income redistribution towards sectors with a relatively high marginal propensity to save. In particular, insofar as increased government or business revenues from price rises of domestically-produced energy are initially unspent, output losses in addition to those induced by OPEC would probably occur. Part III briefly describes the OECD Secretariat's model of the world economy,[1] which has been used to simulate the macroeconomic effects of the 1979/80 oil price rises on Denmark, Finland, Norway and Sweden. After examining the structure of the energy sectors of each of the four countries in Part IV, the simulation results are presented and discussed in Part V. The sensitivity of the results to changed international and domestic assumptions is also considered. In Part VI, the paper concludes that the most pervasive deflationary force on Denmark, Norway and Sweden (and most other industrial countries) would be the transfer of unspent real income to OPEC. For Finland, however, simulated overall output is largely unchanged, because that country's impact real income loss to the Soviet Union (which supplies two-thirds of its oil requirements), is likely to be re-injected virtually immediately as export demand. But for each country, the magnitude of simulated deflation is also shown to depend critically on the assumed degree of unspent revenues arising from domestic energy price rises (especially in Norway), from fiscal drag, and the extent of real wage restraint. Possible induced policy and exchange rate changes, which could be important influences, are not, however, incorporated in the quantitative results.

II. The Macroeconomic Effects of a Real Income Loss Induced by a Rise in the Price of OPEC Oil

This section presents the principles typically used to evaluate the macroeconomic consequences of an oil price rise on industrialised countries that import crude oil and produce some of their own energy. The analysis excludes medium-run effects such as possible changes in the potential GNP growth rate, price-induced increases in energy supply and energy-saving investment, and substitution towards alternative energy sources or towards labour. Rather, it concentrates on the short-term international and domestic redistribution of real income associated with a rise in the price of oil.

---

[1] All Divisions of the OECD Secretariat's Department of Economics and Statistics have contributed to the model's specification, development and implementation. The author is grateful to all colleagues who have helped with quantitative aspects.

Rise in the Price of Imported Oil

For an oil-importing country, a rise in the price of that oil reduces total real national income. The initial income loss, following the terms of trade deterioration, may be measured broadly by the increased cost of imported oil, perhaps expressed as a proportion of total output. As outlined below, the oil price increase *per se* does not, however, necessarily reduce an economy's output; hence, the share of imported oil in total energy requirements is not necessarily a good indicator of the likely extent of deflation.

Such a price increase may be likened to an indirect tax increase, associated with which is a multiplied effect on national income. The multiplier value depends critically on how quickly, if at all, the recipient of the indirect tax increase re-injects the increased revenue back into the economy. According to the Haavelmo-Gelting balanced budget multiplier theorem, equal changes in indirect taxes and government spending have a non-neutral effect on national income, to the extent that the two fiscal instruments have different multiplier values.[1] In a closed economy, output may eventually rise, without any change in the budget deficit.

Extending this principle to the closed world economy, and regarding, for the purposes of exposition, OPEC as the "government' and the rest of the world as the "private sector", it follows that, if the price of imported oil were raised (an "indirect tax" increase), the extent to which output would be reduced would depend crucially on the extent to which the oil-exporting countries bought goods and services from the oil-importing rest of the world. In the extreme case of oil exporters immediately spending all their supplementary income, importing countries' output would eventually be stimulated, if the multiplied (deflationary) effects arising from the oil price increase were more than offset by the multiplied (reflationary) effects of additional export receipts.[2] Within the oil-importing countries, output could be higher with unchanged real income (although a redistribution of real income towards the rest of the world's exporting sectors might have supplementary consequences).[3]

But in practice, OPEC producers typically spend initially only a fraction of their additional oil income. This is particularly true for the sparsely-populated, so-called "low-absorbing" OPEC countries. The private sectors or governments of the oil-importing countries might initially borrow OPEC funds to finance the loss of real income associated with the increase in the oil price. But eventually,

---

[1] Because of first-round savings leakages, an indirect tax multiplier is generally lower than a government non-wage expenditure multiplier; typical values for the Nordic countries' 2-year multipliers are $\frac{1}{2}$–1 and $1\frac{1}{4}$–$1\frac{3}{4}$, respectively, according to OECD (July, 1980). In such cases, equal increases in indirect taxes and government expenditure have an expansionary effect on economic activity.

[2] This depends principally on the relative size of the savings and import leakages of all sectors of the entire economy which are affected by an oil price rise and those of the export sector. Relative multiplier values can only be established empirically.

[3] For example, increased demand for the rest of the world's exports might lead to higher prices, which would partially offset the higher output.

in the absence of full spending by OPEC, it could be expected that spending in the rest of the world would adjust to the lower level of real income. Accordingly, output would fall, relative to what would have happened otherwise.

The amount of OPEC spending on Nordic exports is very limited.[1] How-ever, Finland imports around two-thirds of its crude oil requirements from the Soviet Union. A special bilateral trade agreement between these two countries means that the marginal propensity to spend out of increased Finnish export receipts is nearly unity,[2] compared with a short-term OPEC spending propensity of probably around only one-quarter. Although the remainder of Finland's imported crude oil originates mostly in OPEC countries, a considerable portion of Finland's increased oil bill is, therefore, under present arrangements, likely to be reinjected relatively quickly into the economy. Because of higher export volumes, the deflationary impact of the oil price rise is likely to be considerably less than in countries wholly or largely dependent upon OPEC.

The analysis above is principally couched in terms of real income and spending considerations in the world economy. An alternative approach, the general principles of which are presented by Dorrance (1978), examines the adjustment of nominal financial balances and gives greater prominence to interest rate changes and nominal incomes. This approach leads to the same broad conclusions. In brief, if there is incomplete spending of increased reve-nues by OPEC countries following an oil price rise, then in the face of higher inflationary expectations, domestic financial balances in the rest of the world are likely to be initially reduced—by lower saving, or borrowing—in order to maintain real spending. But after this initial financing period, in which there would be a reversal of the initial upward pressure on interest rates (due to the initial increase in demand for loanable funds), there would be an adjust-ment of real expenditures. Insofar as real financial balances were restored to original levels, the ultimate effect, in the oil-importing countries, would be a lower real income level than would otherwise have been the case. The profile and magnitude of the eventual output decline expected under this approach would depend crucially on the hypothesized lags and elasticity of real spending to real interest rates.

Indigenous Energy Production

A second potential deflationary force may arise if the price of domestically-produced energy is raised in sympathy with a rise in the price of imported oil,

---

[1] The 1979 shares (of total OECD exports) in the OPEC market were as follows: Denmark 0.7 per cent; Finland 0.5 per cent; Norway 0.3 per cent; Sweden 1.8 per cent.

[2] The Soviet Union sells oil to Finland at world prices. Annual protocols on the exchange of goods establish quotas, with a view to balancing a clearing account of bilateral trade. If balance had not been achieved in the previous year, this imbalance would be taken into account in subsequent negotiations. Although agreements are designed to balance the clearing account over a 5-year period, negotiations in fact attempt to balance the account each year. If world oil prices rise, Finland fairly quickly receives additional export orders equal in value to the increased cost of oil.

and all or part of the additional revenue is effectively withdrawn from the economy. Suppose, for example, that the private sector produces oil and raises its price. If the additional revenues were taxed completely and not spent, or if after-tax revenues were not immediately spent on investment or distributed as dividends in that country, then real income would fall in the non-oil private sector.[1] As a consequence, total real private spending would also fall, albeit with a lag, unless savings fell permanently. Where the government itself produces oil or any other form of domestic energy, such as hydro or nuclear electricity, coal or natural gas, and raised its prices "in sympathy" with a rise in world oil prices, then, insofar as the increased revenues were used to reduce the fiscal deficit (or to increase the surplus), real income would be transferred to a sector of the economy with a relatively high propensity to save. These withdrawals of potential expenditure would have associated multiplied deflationary effects on real national product.

The critical factors determining the extent of deflation (or conceivably, reflation) from this source are: the quantity and price of each domestic energy source; whether the additional revenues from total production (including those from exported energy) are sterilised in the event of a price rise; and the redistribution of domestic real income.

Third-country Effects

The mechanisms described above come into play in all countries simultaneously, although to varying extents. The demand for non-oil imports falls by an amount determined principally by income elasticities, and this reduces (export) demand for Nordic products. The effect of this fall could well exceed the increased demand by oil-producing countries. Thus, an additional external deflationary force is likely to be exerted on each Nordic economy.

Wages and Factor Shares

If nominal wages rise, whether because of indexation agreements or an attempt by wage earners to offset the real wage effect of the price rise, the real income shock facing wage-earners is diminished, at least initially, thereby adversely affecting the business and government sectors (insofar as they cut back their expenditure elsewhere in order to pay higher salaries).[2] In the short term, because of money illusion, real private consumption might be boosted because of higher nominal wages. But wage-induced price rises would also act to lower disposable incomes, and would accentuate fiscal drag. Further, if real wages did not fall, firms would probably eventually reduce employment in an attempt to restore profits. In the longer term, these factors could more than offset the

---

[1] If spent abroad, some of the additional revenues may flow back through international trade.

[2] For the government sector, this is considered unlikely in the Nordic countries, but may be the case in some Nordic trading partners.

rise in nominal wages and unemployment benefits, resulting in lower real private consumption.

Private investment, too, would fall, to the extent that it depends on the expected growth of overall output, which declines incrementally following the real income loss. Any downward impact on investment would be accentuated as real wage resistance strengthened, if, as would seem likely, lower business profits lead to postponement or cancellation of investment projects. If higher wage costs were passed into higher export prices to mitigate a profit squeeze, there would be a deterioration in international competitiveness, harming export volumes. Insofar as the exchange rate devalues, any accompanying additional loss of real income would adversely affect both real consumption and real investment. The effect would be similar if the nominal OPEC oil price rose in the face of higher OECD export prices for manufactures. The likely effects of interest rate changes would seem ambiguous.[1] On the basis of the above considerations, the eventual impact of oil-price-induced higher nominal wages would be to add further deflation.[2]

Non-oil Terms of Trade

The overall deflationary impact of an oil-price-induced real income loss might be reduced for an oil-importing country if its non-oil terms of trade improved after an oil price shock.[3] The price of certain raw materials and commodities, especially food and metals, might fall considerably with an oil-price-induced decline in overall world demand, especially in the short term. But export prices of manufactures or semi-processed products are unlikely to fall by as much, especially if foreign demand was relatively price and income inelastic or if they were energy-intensive.

Given the composition of the Nordic countries' trade, and that the market

---

[1] Interest rates would rise if nominal money supply targets were adhered to, and if real domestic spending propensities were unaltered. On the other hand, increased world savings, associated with the OPEC surplus not fully offset by increased planned investment in the short term, would tend to reduce money demand and hence interest rates. (Eventually, equilibrium of *ex ante* world savings and investment would be restored.) The direction of change of real interest rates, and, hence, the effect on investment, is therefore ambiguous.

[2] Possible policy reactions have been ignored. Following the 1973/74 oil price rises, there was an increase in the four Nordic countries of the growth of compensation of employees, including a strong element of wage drift during 1974. But because of fiscal policy changes and the implementation of incomes policies at various stages through 1974/76, it is not possible to estimate with any degree of certainty whether the oil-induced wage increases added to deflation.

[3] By non-oil terms of trade is meant the ratio of non-oil export prices to non-oil import prices. Insofar as the composition of imports and exports differs and each changes over time, estimates of gains or losses in the terms of trade can vary somewhat according to how they are measured. Inter-country comparisons are hazardous when the improvement of the output price of an exported product due to the passthrough of the increased cost of oil inputs leads to a gain in the "non-oil" terms of trade. This is especially so during a period when energy inputs per unit of output are changing, i.e. quantity as well as price changes may lead to terms of trade changes. Moreover, terms of trade changes induce changes in real activity which in turn may affect relative prices, so the final terms of trade effect depends on the (arbitrary) time period chosen.

for certain products is oligopolistic, an improvement in the non-oil terms of trade could be expected after an oil price rise, at least initially.[1] But the extent of any ultimate improvement would depend on the degree to which Nordic exporters are price-setters, which in turn depends on the change in relative unit labour costs and final product prices *vis-à-vis* principal competing trading partners, and upon any induced exchange rate changes. An offsetting factor would be the extent to which countries which also incurred a non-oil terms of trade loss further cut back their demand for Nordic countries' products.

Final Price Response

The direct effects of a rise in the price of imported oil would soon show in import and consumer price indices. Indirect effects, coming through higher prices for domestically-produced and consumed energy, and a higher cost for labour and capital, would add to the final price response. Mitigating factors would be the extent to which the fall in aggregate demand reduced overall prices, and the extent to which enterprises did not pass higher costs onto final product prices. For a net oil-importing country, the inflationary impact might be magnified if the exchange rate devalued, although the induced improvement in competitiveness would offset the negative output effect, which in turn could affect final prices, especially if there were capacity constraints. The reverse exchange rate repercussions on inflation could be true for a net oil exporter such as Norway. Because the lags in these processes are relatively long, however, they have not been considered in the simulation exercise reported in Part V.

Policy Response

The demand-deflationary and cost-inflatory responses outlined above have been considered in the assumed absence of a policy response by governments and central banks of both the oil-importing and oil-exporting countries. A government of an oil-importing country might attempt to reduce the inflationary impact by reducing indirect taxes or by attempting to influence wage settlements.[2] It might offset the demand reduction and adverse employment effects through increasing subsidies and other transfers, or by cutting income taxes of households or businesses, or by increasing government expenditure elsewhere. Monetary policy could be accommodating; or, in the face of increased inflationary expectations, or for external reasons, monetary targets could be unchanged, with a resultant upward pressure on real interest rates. If a country's competitive position deteriorated, the exchange rate might

---

[1] Indeed, after the 1973/74 oil price rise, each Nordic country's non-oil terms of trade (measured by the ratio of unit values of exports divided by unit values of non-oil imports) improved between 1973 and 1975.

[2] For example, energy prices were recently removed from the Danish wage-regulating price index. Another possibility is for income tax reductions to be offered in return for wage moderation.

be altered by the authorities. Finally, the oil-exporting countries might adjust production consistent with desired oil revenue to ensure that the new real oil price "stuck", or they might alter their spending propensities in the face of increased revenues. All these, and other possible policy responses, are important in assessing the likely overall macroeconomic impact. However, in the simulations that follow, a no-policy-change assumption is generally adopted as outlined below.

III. The Model Used

The simulations reported in Part V are based on the September 1980 version of the OECD Secretariat's world economic model, INTERLINK. Earlier versions of this model have been described elsewhere; see OECD (January, 1979 and July, 1980). This section briefly outlines the present structure of the model, highlighting features which are particularly relevant for the assessment of a rise in world oil prices.

The model consists of twenty-three OECD country models, eight non-OECD regional models and a world trade block which links the OECD country models with each other and with the non-OECD regions through merchandise and service trade volumes and prices. In oil price simulations, a four-way commodity split of merchandise trade enables the likely effects of additional deflation coming from third countries' reduced import levels to be captured. The non-OECD regions are assumed to spend their incremental export receipts with a lag, which is taken to be short for the Soviet block and developing countries, but long for the OPEC countries.[1] Each country model contains over one hundred equations, but the basic principle of drawing on empirical results of larger national models in order to mimic their key simulation properties has been retained; see OECD (July, 1980). Rather than generating baseline forecasts itself, INTERLINK locks on to historical data and OECD forecasts, through the generation of add-factors.

Simulated private consumption and investment are determined principally by changes in simulated current and lagged real household income, and changes in real output growth, respectively. Given a hypothesized oil price rise, each of these variables initially falls to the extent that higher simulated import prices are passed into simulated consumption and investment deflators. Government wage payments and transfers respond, inter alia, to the degree of indexation in each country, but nominal government investment is generally assumed to be given.

A set of appropriation accounts in each country model ensures that simulated net lending of the household, business and government sectors equals

---

[1] The OPEC block is subdivided into high and low absorbers, with particularly long spending lags for the latter.

the simulated foreign balance. Following an oil price rise, the government sector receives increased tax receipts, reflecting a simulated increase in the nominal income base. In the cases of Denmark, Finland and Sweden, simulated incremental government wage payments and net transfers (especially unemployment benefits) may be sufficient to offset simulated fiscal drag on household incomes, so that the government may initially bear some of the simulated foreign deficit. But for Norway, the effective overall business marginal tax rate is set considerably higher than for the other three countries, taking into account royalties and other special taxes in the oil sector. These simulated "windfall" tax receipts from the oil sector exert a strong positive influence on the government's simulated financial surplus.

In each country, simulated household saving ratios fall initially, as simulated real spending adjusts with a lag to simulated real income, entailing a deterioration in household financial balances. The enterprise sector's simulated financial position also deteriorates, except in Norway. But because this is presumed to occur entirely in the oil-producing sector, increased simulated business savings are not presumed to affect domestic investment. This assumption may result in a failure to capture increased investment in oil-related industries. A possible offset to this, which also is not modelled, is the extent to which the oil sector's exports induce an appreciation of the exchange rate and depress investment in non-oil activities.

Turning now to simulated prices and wages, the latter respond to higher simulated consumer prices and export prices of the manufacturing sector, with adjustment presumed to take up to eighteen months. Insofar as expected prices are based on past price changes, the consumer price term could be interpreted as the influence of expected price changes on wages. Consumer prices react virtually immediately to a change in the price of imported oil. The passing-through of non-oil import prices is assumed to take considerably longer —up to eighteen months. The price of domestically-produced and consumed energy is assumed to increase by around a half of the price rise of imported oil, with a lag of three years. The pass-through factor is based, in part, on the 1973/78 experience of other energy prices, presented in Table 2. On the assumption that a fall in aggregate demand typically has only a minimal effect on prices and wages in the short term, that potential mitigating factor of the final price response is excluded from the simulations reported below.

IV. Characteristics of the Energy Sector in the Four Countries

This section briefly reviews total energy requirements (TER) for domestic use, and energy pricing and conservation policies in each country. Table 1 shows TER in 1978 by four different sources. Denmark imports nearly all of its domestic energy requirements; 80 per cent of imported TER is crude oil and petroleum products. Finland and Sweden both depend on imported oil (including products) for nearly 60 per cent of TER, though from different countries of origin.[1] Imported coal supplies 15–20 per cent of TER in Denmark and Finland. Even without North Sea oil, Norway is well supplied with domestically-produced energy. Although total electricity generation is not as great as that of Sweden (the latter having the highest nuclear power/TER ratio of all OECD countries in 1978), hydro-electric power provides some 55 per cent of Norway's TER. Production of crude oil and natural gas was nearly twice domestic energy needs in 1979.

Each country faces world prices for imported crude oil and petroleum products. Insofar as Sweden buys proportionately more refined products on the spot market, where prices were very high in 1979, it was temporarily faced with a higher average dollar import price than the other countries. Exchange rate movements until mid-1980 dampened the local currency price rise for each country except Denmark.

From Table 2 it can be seen that electricity prices to industry were considerably lower in Norway than in its Nordic partners in 1978. This was particularly the case in certain electricity-intensive industries. In none of the countries have electricity prices risen as fast as the untaxed component of petroleum-based products. And in all cases, the cost of electricity to industry is considerably below that to consumers. The prices of domestically-produced solid fuels, which are quite important energy sources in Sweden and Finland, have generally risen in line with production costs. Traded coal, which is an input for electricity generation in Denmark and Finland, has risen in price since 1973 by an amount considerably less than that of crude oil. As a consequence, in Denmark, coal represented 66 per cent of the fuel in electricity plants in 1979 (and was still increasing rapidly during 1980) compared with 42 per cent in 1973. Finally, Norwegian oil and gas, nearly all of which has been exported so far, are sold at prices comparable to those of other suppliers (with a 12-month adjustment lag in the case of gas). In the simulations presented below, it is assumed that export prices of North Sea crude oil adjust immediately to any rise in the OPEC export price.

Each Scandinavian country encourages conservation and fuel-switching. Measures such as subsidies and tax incentives for better insulation, maximum heating limits in public buildings, changed building codes, maximum road speed limits, changes in indirect taxes on energy products, and so on, have contributed to the lowering of the energy per unit of GDP ratio[2] from 0.71 in 1973 to 0.68 in 1978 for the four countries as a group. Emphasis on reducing

---

[1] The crude oil/petroleum-based products mix is also different. Finland relies heavily on the USSR, and in 1978, petroleum products accounted for only 17 per cent of imported energy requirements. In contrast, Sweden is reliant mainly on OPEC oil; and in 1978, petroleum products were 42 per cent of imported energy requirements (most of which were bought on the spot market).
[2] Total primary energy (in tons of oil equivalent) per thousand 1970 U.S. $.

Table 1. *Total primary energy by source, 1978*[a]

Million tons of oil equivalent (mtoe)

	Denmark			Finland			Norway			Sweden		
	Indigenous production	Imports	Total energy requirements (TER)	Indigenous production	Imports	Total energy requirements (TER)	Indigenous production	Imports	Total energy requirements (TER)	Indigenous production	Imports	Total energy requirements (TER)
Crude oil, NGL[b] and gas	0.44	7.84	8.09	—	11.51	12.05	29.16	7.65	9.65	—	16.39	15.45
% of total	63.7	34.2	39.5	—	63.2	50.1	70.0	71.9	43.4	—	53.6	31.4
Petroleum products	—	10.74	8.17	—	3.02	1.11	—	2.38	-0.14	—	12.71	10.08
% of total	—	46.8	39.9	—	16.6	4.6	—	22.4	-0.6	—	41.5	20.5
Solid fuels	0.24	3.99	3.90	3.63	3.54	7.52	0.26	0.53	0.77	3.17	1.29	4.66
% of total	34.7	17.4	19.0	52.8	19.4	31.3	0.6	5.0	3.5	14.2	4.2	9.5
Electricity[c]	0.01	0.38	0.33	3.24	0.13	3.35	12.23	0.07	11.94	19.14	0.21	19.05
% of total	1.5	1.7	1.6	47.2	0.7	13.9	29.4	0.7	53.7	85.8	0.7	38.7
Total	0.68	22.95	20.48	6.86	18.21	24.03	41.66	10.64	22.22	22.32	30.60	49.24
Memorandum items												
% of imported oil, gas and refined products in TER												
1973	107.5			61.6			60.0			63.9		
1978	90.7			57.0			45.1			59.1		

[a] *Source:* "Energy Balances in OECD Countries, 1974/78", Paris, 1980. For any given country, the difference between the sum of indigenous production and imports, and total energy requirements gives exports (including marine bunkers) and net stock changes.
[b] Natural gas liquids.
[c] Nuclear, hydro and geo-thermal.

Table 2. *Prices of selected final energy products*[a]

	Denmark			Finland			Norway			Sweden		
	1973	1978	Per cent change	1973	1978	Per cent change	1973	1978	Per cent change	1973	1978	Per cent change
Electricity (U.S. cents/KWh)												
Domestic use[b]	2.57	5.77	125	2.46	5.09	107	2.99	6.49	117	2.21	4.11	86
Industrial use[b]	1.70	3.76	121	1.69	3.84	127	1.14	1.96	72	0.96	2.78	190
Coal												
Domestic use[c] ($/ton)	38	69.9	84	n.a.	n.a.	—	71	154.1	117	n.a.	n.a.	n.a.
Industrial use[d] ($/10⁷ Kcal)	n.a.	n.a.	—	n.a.	n.a.	—	46.0	95	106	3.10	55.3	78
Petroleum products												
Gasoline[e] (cents/litre)	24	47.2	97	20.6	49.2	139	24.7	49.6	101	22.0	36.8	67
Domestic gas/diesel oil[f] (cents/litre)	5.2	17.7	240	n.a.	14.4	—	5.6	17	204	4.4	13.1	198
Industrial heavy fuel oil[g] ($/metric ton)	22.6	124.6	451	36.9	99.5	170	28.6	128.2	348	32	98.7	208
Memorandum items												
Consumer price index (annual average)	100	161.9	62	100	189.0	89	100	157.3	57	100	168.2	68
Exchange rate (U.S. cents/ unit of local currency)	15.92	17.31	8.7	25.64	24.83	-3.2	16.72	19.46	16.4	21.93	21.42	-2.3

[a] *Source:* "Energy Statistics", OECD, Paris, 1979, together with country submissions to the International Energy Agency for certain series. Prices, including exchange rates, are those prevailing at 1st January each year.
[b] For annual consumption of 5 000 KWh per annum (domestic) and 15 GWh per annum (industrial). Includes direct taxes in both cases for each country.
[c] Low volatile coal, 500 kg lots.   [e] Supergrade gasoline (including taxes).   [g] 500 tons per annum.
[d] Washed steam coal, 0–10 mm.   [f] 5 000 litre lots.

the volume of oil imports through price-induced inter-fuel substitution has resulted in an even greater fall in the imported oil/GDP ratio.[1]

A crucial question in the special case of Norway is the extent of spending of additional oil and gas revenues from an OPEC-induced oil price rise. The marginal tax rate of increased earnings of companies operating in the North Sea, so far largely foreign-owned, is assumed to be 80 per cent.[2] Oil tax revenue was estimated to account for 18 per cent of central government revenue in 1980. It is assumed that these revenues are used to reduce the government's budget deficit. The counterpart to this would be increased foreign exchange revenues or a reduction of government debt funded from abroad. The oil companies' additional after-tax income is assumed to be transferred abroad, either for foreign investment or to share-holders in the firms' home country. Because all of the increased oil revenue is transferred either to the government, which by assumption does not spend it, or abroad, an oil price rise is likely to be deflationary under such assumptions. Another important deflationary force is the decline in freight earnings received by Norwegian shipping companies, which are especially active in the spot freight market, transporting oil to various oil-importing countries. Following an oil price rise, this element of service exports (which represents around 20 per cent of total Norwegian export receipts) could be expected to fall markedly. Major uncertainties for Norway include whether or not the increased income of oil companies generates new oil investment,[3] the extent to which overall business confidence induces additional investment in sectors partially related or unrelated to the oil industry, the economic rent gained by energy-intensive Norwegian industries,[4] whether the government in fact spends some of the additional revenues or reduces taxation, and whether the foreign balance improvement induces strength in the krone that would be viewed as undesirable for traditional exports. A broader uncertainty is the extent to which successful real wage claims in the oil sector are imitated in the mainland economy.

V. Simulated Effect of the 1979/80 Oil Price Rise on the Four Nordic Countries

This section first outlines the main assumptions underlying the results. It then examines how the simulated effects could differ if changes were made in some

---

[1] From 0.45 in 1973 to 0.38 in 1978 (same units as for total primary energy). Source: International Energy Agency.

[2] Fields with major Norwegian-owned interests are not yet in production. For details of the effects of the 1975 tax package; see Kemp and Crichton (1979) and *OECD Economic Survey* (January, 1980).

[3] A higher oil price would make profitable oil fields which were previously marginal. Whereas the speed of development of such fields is open to conjecture, it has been assumed that short-term planning and capacity constraints, together with the Norwegian government's procedure for granting licensces, would prevent any additional investment being made over a 2–3 year time horizon.

[4] In the event of an oil price rise. Norwegian energy-intensive industries that rely on cheap domestic hydro-electricity are place in an advantageous position *vis-à-vis* corresponding

Table 3. *Simulated impact of oil price rise on GDP, consumer prices and current balances*

The July 1980 OECD *Economic Outlook* forecasts terminated in mid-1981. For this reason, 1981 baseline levels are not shown

	Real GDP (billions of local currency, 1978 prices)			Consumer prices[a] (Index: 1978 = 100)			Current balance (billions of local currency units)		
	1979	1980	1981	1979	1980	1981	1979	1980	1981
Denmark									
Without oil price rise	322.2	326.7	—	107.1	115.9	—	− 10.9	− 12.9	—
With oil price rise	319.0	317.3	—	109.6	123.9	—	− 15.6	− 20.3	—
Percentage difference[b]	− 1.0	− 2.9	− 3.7	2.3	6.9	11.6	− 4.7	− 7.4	− 6.5
Finland									
Without oil price rise	148.6	157.3	—	105.5	111.1	—	− 0.0	− 3.6	—
With oil price rise	148.5	157.4	—	107.5	119.5	—	− 1.0	− 6.0	—
Percentage difference[b]	− 0.0	+ 0.0	+ 0.0	1.9	7.6	12.6	− 1.0	− 2.4	− 2.4
Norway									
Without oil price rise	220.1	232.9	—	103.0	107.8	—	− 9.6	− 2.4	—
With oil price rise	219.0	229.0	—	104.7	114.4	—	− 5.8	+ 7.8	—
Percentage difference[b]	− 0.6	− 1.7	− 3.6	1.7	6.1	11.3	+ 3.8	+ 10.2	+ 11.6
Sweden									
Without oil price rise	413.3	432.7	—	105.0	114.2	—	− 6.1	− 12.1	—
With oil price rise	409.5	421.6	—	107.2	121.3	—	− 10.8	− 18.9	—
Percentage difference[b]	− 0.9	− 2.4	− 2.9	2.1	6.2	9.9	− 4.7	− 6.8	− 5.3

[a] Implicit deflator for private consumption.
[b] Except for the foreign balance, which is in level form.

key assumptions. The results are hypothetical to the extent that no attempt has been made to adjust the magnitude or profile in order to account for known policy changes that occurred up to mid-1980. Moreover, it may well be that the private sector's short-term response was different from assumed, because of changed confidence of consumers and enterprises.

Using baseline OECD forecasts,[1] the simulated effects on GDP, consumer prices and current account balance levels are shown for each country in Table 3. It is assumed that, in the absence of the oil price rise, the real price of oil would have been constant,[2] and that OPEC import volumes of goods and services would have increased by around 5 per cent.[3] *Ex post*, around one

energy-intensive industries abroad that rely on crude oil. The profits of such industries rose sharply in 1979.

[1] The INTERLINK simulations generally result in linear increments. The choice of baseline (forecast or actual values) is therefore unimportant.

[2] The rate of growth of the dollar price of imported oil was set at around 6 per cent p.a. on average, the rate at which industrial countries' export prices of manufactured products may have increased in the absence of the oil price rise.

[3] In line with the growth rate generally expected in late 1978. However, as OPEC import volume growth was weaker than 5 per cent through 1979, in part because of an important import cutback by Iran, it is assumed that, even in the absence of the oil price rise, import volume growth through 1979 would not have been weaker than the actual outturn.

quarter of OPEC's simulated cumulative export receipts are spent on imports by 1981. Local currency oil import prices for the four Nordic countries, as well as those for all other OECD countries, are assumed to increase by around 150 per cent by 1981.[1] An unchanged real oil price is assumed as from mid-1980.

Other key assumptions are:

(i) Monetary policy is assumed to accommodate changes in nominal GDP, without affecting real interest rates. The flow of OPEC revenues into the industrial countries' banking systems, and other capital transfers which could affect interest rates, are not explicitly modelled. Exchange rates are held fixed from mid-1980.

(ii) Discretionary elements of fiscal policy such as tax rates and non-wage expenditure are exogenous. Government wage payments, transfers and tax receipts respond endogenously.

(iii) A moderate domestic wage/price response, including that of domestically-produced energy prices, is assumed, as outlined in Part III. International commodity prices do not respond to changed demand conditions in industrial countries.

The overall simulated output effect for the four countries in aggregate, as measured by a GDP decline of some 3 per cent relative to what would have occurred in the absence of the oil price rise (see "Standard Case" of Table 4), results from a combination of changed domestic and foreign demand, each effect differing appreciably amongst the four countries. Despite Finland having by far the highest TER/GDP ratio of the four countries,[2] it has the lowest simulated output response to the assumed oil price shock. This is largely due to foreign demand from the Soviet Union being maintained and domestic demand (investment in particular) holding firm as a consequence. Norway's output change is simulated to be of the same order of magnitude as Denmark's. Although the former country's imported oil/GDP ratio is the lowest of the four, it nonetheless has a large negative simulated contribution to its real foreign balance (although of course, the nominal foreign balance improves substantially). An important factor is that, of the four countries, Norway has the lowest share in the OPEC market. Simulated real freight earnings also contribute considerably. Norway's simulated oil export volumes do not fall much, for OPEC has been assumed to be the marginal crude oil supplier. Despite Denmark's being the most energy-efficient economy of the four,[3] its

---

[1] The price of imported oil includes that of petroleum products. In 1979, the local currency price increase was around 50 per cent on average, being a little stronger than this in Denmark and Sweden. The 1980 increase is assumed to be around 45 per cent and that of 1981 to be around 10 per cent. Differences between countries mainly reflect different crude oil/petroleum product weights and differing exchange rate developments to mid-1980.

[2] In 1978, the TER/real GDP ratios (in mtoe per 1970 U.S. dollar GNP) were: Denmark 0.51, Finland 0.88, Norway 0.69, Sweden 0.72.

[3] As measured by the TER/GDP ratio. Two reasons are that the structure of Danish output is by far the least energy-intensive and that the climate is slightly warmer.

negative simulated real output effect is one of the largest, the main reasons being its higher dependence on OPEC oil for total energy requirements and the higher local currency increase in the price of imported oil until mid-1980. Because of its greater proportion of exports to OPEC, the simulated output loss in Sweden is less than that in Denmark and Norway.

Simulated consumer prices in 1981 are, on average, some 11 per cent higher than would otherwise have been the case. This effect excludes the higher indirect taxes and price freezes which were important influences on price developments in Denmark, Norway and Sweden during 1979 and 1980. GDP deflators are not shown in the tables, but they are simulated to change by about three-quarters of this amount, except in Norway, where the overall terms-of-trade improvement from the oil price rise results in the simulated deflator being some 16 per cent higher by 1981.

The main factors influencing the magnitude of the simulated current account balances are the magnitude of the oil price increase, the importance of net oil imports (or exports) in each economy, the response in cutting back oil imports as a consequence of an oil price rise,[1] the simulated effect of non-oil trade (including that of services) in the face of changed world demand, and trade shares.

The sensitivity of the simulated effects of the oil price rise to certain key assumptions is shown for the four economies in aggregate in Table 4. In the absence of any nominal wage response in the OECD economy to higher oil-induced domestic prices, the simulation results suggest an output loss by 1981 some 2 per cent less, with the impact on inflation halved. When wages do respond, simulated consumer expenditure is initially marginally higher, as higher nominal wages, together with a simulated endogenous weakening of saving propensities, offset higher consumer prices. But with a greater squeeze on the non-oil company sector's profits, and households financial positions adversely affected by greater fiscal drag, the eventual simulated impact is demand-reducing, especially for private investment. With the moderate wage/price spiral assumed, some additional inflation is transmitted internationally, accentuating the eventual simulated deflationary effect.

The Scandinavian countries generally have a high degree of progressivity in their tax systems. The short-term elasticity of household tax payments to the household taxable income base (under unchanged tax scales) may be as high as 1.7, but over a 3-year period, governments generally raise thresholds or reduce rates to offset fiscal drag. If the assumed 3-year elasticity of 1.15 were reduced to 1.0 (i.e. full indexation of the tax system), the simulated deflationary impact of the oil price rise would be reduced by more than half a per cent by 1981.

Without the assumed eventual 50 per cent passthrough of higher imported

---

[1] Income and price elasticities are roughly the same for each country.

Table 4. *Sensitivity of oil price simulation to various assumptions. Aggregate of four Nordic countries*

Percentage difference from baseline level, except for current account balance, which is in U.S. dollars

	Real GDP			Consumer prices			Current balance (U.S. $ billion)		
	1979	1980	1981	1979	1980	1981	1979	1980	1981
Standard case[a]	−0.7	−2.0	−2.8	2.0	6.5	11.0	−1.6	−1.4	−0.9
Domestic factors									
(1) Standard case without wage/price feedback	−0.7	−1.5	−0.7	1.6	4.2	5.3	−1.6	−1.7	−1.4
(2) Standard case without an adjustment of domestic energy prices	−0.6	−1.6	−1.6	1.8	5.6	9.1	−1.6	−1.5	−0.5
(3) Standard case without fiscal drag	−0.5	−1.9	−2.2	1.9	6.2	10.1	−1.2	−1.2	+0.3
International factors									
(4) Standard case without OPEC res) pending	−0.8	−2.9	−5.1	2.0	6.5	11.0	−1.6	−2.5	−2.5

[a] For GDP and consumer prices, the figures are a weighted average of the Table 3 results for the individual countries. Constant 1979 U.S. dollar GDP and private consumption weights respectively were used.

oil prices to domestically-produced and used energy (the proceeds of which are not spent), the simulated inflation impact of 11 per cent by 1981 would be reduced to around 9 per cent; there is a corresponding reduction in the simulated real GDP loss, by 1¼ per cent, arising from the lower incremental decline in private sector real income.

If the modest assumed OPEC spending were not to occur, then Denmark, Norway and Sweden in particular, as well as the principal Nordic trading partners in the Western world, would be adversely affected. Because Sweden has the largest share in the OPEC market of the four countries, it would be the most critically affected by an absence of OPEC spending. Simulated GDP for the four countries as a group would fall by almost twice as much, a major change occurring in the additional simulated deflation between 1980 and 1981. It is estimated that over one-third of the additional 2¼ per cent deflation originates from lower demand by principal Nordic trading partners. In the standard case, no additional deflation of output occurs in the simulation by the end of 1981—increased OPEC spending, and an assumed constant real oil price, is simulated to have an eventual reflationary impact on the oil-importing world. And in the case without an assumed wage/price response, the assumed OPEC spending has a significant simulated reflationary impact on output by 1981.

VI. Conclusions

This study has attempted to quantify some of the major factors affecting output and inflation in four Nordic economies, arising from the 1979/80 rise in OPEC oil export prices. By 1981, simulated real GDP and consumer prices in the four countries in aggregate are 3 per cent lower and 11 per cent higher respectively than otherwise would have been the case. But the diverse results between countries reflect in particular the differing marginal spending propensities of the eventual recipients of increased oil revenues. The high OPEC short-term marginal propensity to save exerts a strong deflationary influence on Denmark, Sweden and Norway, despite the latter country's favourable energy balance. Finland has the lowest simulated output loss of the four countries, because the Soviet Union is assumed to spend its increased oil revenues from Finland very quickly. Under the stringent assumptions that 80 per cent of Norway's North Sea oil revenues are taxed and not spent by the government and that the remaining 20 per cent is redistributed abroad by the foreign companies which receive the additional proceeds, simulated GDP in Norway also falls. Norway's very limited share of total OPEC imports is also an important factor leading to this result.

Variations in reactions of domestic sectors could alter the magnitude and profile of the results. In the short-term, a nominal wage response may boost overall expenditure, because households' marginal propensity to save is assumed to be lower than that of businesses. But without the assumed moderate wage/price spiral, the simulated incremental output effect after three years is reduced by some 2 per cent. The reduction of the assumed fiscal drag on household incomes also reduces the simulated output decrement somewhat. The upward adjustment of other domestic energy prices in sympathy with an oil price rise exerts an additional thrust to the simulated output decline, especially in the third year, when its impact on consumer prices, and, hence on real household incomes, is simulated to have its strongest effect.

If OPEC were not to spend any of its additional revenue, rather than the assumed fraction of about a quarter by 1981, the simulated output reduction by 1981 is raised by over 2 per cent, of which a significant portion is estimated to come from the lower demand of principal trading partners. For open economies, the ultimate output reduction is critically dependent not only on the direct effects of low OPEC spending in the short term, but also on indirect effects operating through world demand in general.

Attention has been focussed on some of the key issues associated with a rise in OPEC oil prices. Other factors not explicitly considered here, such as exchange rate reactions and changes in relative competitive positions, changes in monetary, fiscal, energy, and prices and incomes policies; international capital flows; interest rate effects; and the effect of expectations on price and income determination, could well alter these results. The margin of uncertainty is greatest in the case of Norway.[1] A sharper disaggregation of the oil and non-oil sectors would isolate the diverse macroeconomic effects in the two sectors. The effect of the (excluded) medium-term changes in the distribution of factors of production (oil and non-oil energy, capital and labour) might well be quantitatively important by the third year.[2] All of the above caveats could be areas of further research. But it is unlikely that such refinements would alter the proposition that the transfer of real income to a world zone or a domestic sector with a much lower spending propensity is the most pervasive force in determining the extent of output reduction in the Nordic countries.

References

Corden, W. M.: *Inflation, exchange rates and the world economy*. Clarendon Press, Oxford, 1976.

Eckstein, Otto: The great recession, with a postscript on stagflation. *Data Resources Series*, vol. 3. North Holland Publishing Company, New York, 1978.

Economic Council of Denmark: *Dansk økonomi*. Copenhagen, May 1979.

Dorrance, G. S.: *National monetary and financial analysis*. Macmillan Press Ltd., London, 1978.

Fried, Edward R. & Schulze, Charles L. (eds.): *Higher oil prices and the world economy, the adjustment problem*, The Brookings Institution, Washington, D.C., 1975.

Johnson, K. & Klein, L.: Stability in the international economy: the LINK experience. In *International aspects of stabilisation policies*. The Federal Reserve Bank of Boston, Conference Series, no. 12, Boston, 1974.

Kemp, A. G. & Crichton, D.: The North Sea oil taxation in Norway. *Energy Economics 1* (1), 1979.

Ministry of Finance, Annex to "Revidert nasjonal budsjett 1980" (Revised National Budget), Oslo, Norway, April 1980.

Mork, K. A. & Hall, R. A.: Energy prices, inflation and recession, 1974–75. *The Energy Journal*, July 1978.

OECD Economic Outlook 27, Special Section "The Impact of Oil on the World Economy", OECD, Paris, July 1980.

OECD Economic Survey, "Norway", OECD, Paris, January 1980.

OECD, "The OECD International Linkage Model", *Economic Outlook*, Occasional Studies, OECD, Paris, January 1979.

OECD, "Fiscal Policy Simulations with the OECD International Linkage Model", *Economic Outlook*, Occasional Studies, OECD, Paris, July 1980.

Okun, Arthur M.: A postmortem of the 1974 recession. *Brookings Papers on Economic Activity 1*, 1975.

---

[1] See caveats in the final paragraph of Part IV.

[2] An increase in the price of imported oil would increase the *potential* output growth warranted by a given energy price/volume mix, but it is debatable whether *actual* output growth would increase, especially if energy-saving investment is insufficient to offset cutbacks in other investment.

OPEC RESPENDING AND THE ECONOMIC IMPACT OF AN INCREASE IN THE PRICE OF OIL

Jan F. R. Fabritius and Christian Ettrup Petersen*

The Economic Council, Copenhagen, Denmark

Abstract

An empirical measure of OPEC respending is developed in this paper and used to compare the situation in 1979/80 with that in 1973/74. A world trade model system is used to simulate the impact of an increase in the price of oil on the industrial countries, under various assumptions about OPEC respending. The simulation results suggest that the degree of respending is crucial for the impact of the oil price increase on the industrial economies and that Europe is more sensitive to variations in OPEC respending than the USA.

I. Introduction

After two major oil price hikes in less than a decade, the price of oil has risen to a level more than ten times the pre-1973 price. The industrial countries have experienced a slowdown in economic growth, temporary recessions and, at times, huge balance-of-payments deficits *vis-à-vis* the oil-exporting countries. The transfers to OPEC in payment for more expensive oil imports have not constituted a complete drain on the reserves of the industrial countries; a portion of the extra oil revenues have returned in the form of OPEC import demand, investment, financial "recycling", etc.

The extent to which OPEC respends its additional oil export earnings by buying goods and services from the industrial countries is crucial to the overall impact of an increase in the price of oil.

The OPEC balance of payments development in the 1970s and some of the factors which influence OPEC respending decisions are summarized in Section II. An empirical measure of the degree of respending is developed and used to compare the first and second waves of oil price increases. The analysis indicates that OPEC respending was considerably lower in 1979/80 than in 1973/74.

* We would like to express our gratitude to our friends at the OECD Econometric Unit, notably Lee Samuelson and Fransiscus Meyer-zu-Schlochtern, for their kind cooperation in making data and computer software available for transfer to RECKU, the computer center at the University of Copenhagen, where the model simulations have been carried out. The development of the model was the subject of Christian Ettrup Petersen's graduate thesis, University of Copenhagen (1980). Similarly, we would like to thank our friends and collegues at The Economic Council, notably Jesper Jespersen, Dan Knudsen, Erik Steen Sørensen and Torben Visholm, for useful comments, suggestions and discussions on the article and the subject in general. Financial support from the Danish Social Science Research Council (grant no. 514-20403) is gratefully acknowledged.

The geographical distribution of OPEC imports from the OECD countries and the terms of trade losses from an oil price increase are examined in Section III. Section IV contains an outline of a world trade model system (CRISIS) which is subsequently used to simulate the impact of an oil price increase. The simulation results under various respending assumptions are summarized in Section V. Some conclusions are drawn in Section VI.

The results suggest that the impact of an increase in the price of oil on the OECD countries, assuming unchanged economic policy, is rather small and that OPEC respending has a significant influence on the overall effect. However, a more noteworthy result may be that the geographical distribution of the simulated impact indicates that the European countries are more sensitive to changes in OPEC respending than the USA.

Hence, it can be concluded that although the two oil price hikes in 1973/74 and 1979/80 are comparable as regards the external shock to the OECD countries—about 2 % of GDP in terms-of-trade losses in both cases—lower OPEC respending in 1979/80 is likely to cause a relatively deeper recession in Europe than in 1973/74—conditional, of course, on changes in economic policy.

According to these simulations, the effects of an oil price increase are smaller than in comparable studies previously published by the OECD. This can largely be traced to differences concerning the concept of unchanged economic policy. In other words, the politico-economic reactions of the industrial world and the pursuit of nationalistic balance-of-payments and inflation targets may have been important factors underlying the slowdown in economic growth during the 1970s—perhaps as important as the oil price increases themselves.

II. OPEC: Measuring Respending

The trends of OPEC balance of payments since 1973 are outlined in Table 1. Changes in the balance of trade can be decomposed into changes in quantity (measured at 1973 prices) and price changes.

As indicated by Table 1, the growing export revenues to OPEC countries during the 1970s are almost exclusively due to price increases, whereas the volume of oil exports has hovered around the same level throughout the period 1973–80. As could be expected, the increased dollar value of OPEC imports is the result of price as well as quantity changes. The OPEC trade balance surplus of $167 billion for 1980 (estimated in early 1981) is the net result of a $185 billion gain due to the terms-of-trade improvement from 1973 to 1980 and quantity changes involving a revenue loss of $18 billion. The current balance surplus is smaller—estimated at $114 billion—owing to a net expenditure on services, interest payments and transfers.[1]

---

[1] Since net interest payments can be expected to show a surplus for most OPEC countries due to the accumulation of foreign assets, this implies that repatriated interest payments do not even suffice to cover the need for domestic use of imported services.

Table 1. *OPEC balance of payments, price and quantity changes, 1973–80*

$ billions	1973	1974	1975	1976	1977	1978	1979	1980
Exports								
Quantity at 1973 prices	—	42	38	43	44	43	43	39
Value of price change since 1973[a]	—	74	69	90	101	103	169	266
Current value	42	116	107	133	145	146	212	305
Imports								
Quantity at 1973 prices	—	29	40	45	52	64	47	57
Value of price change since 1973[a]	—	10	18	23	32	40	55	81
Current value	21	39	58	68	84	104	102	138
Trade balance surplus								
Quantity at 1973 prices	—	13	− 2	− 2	− 8	− 21	− 4	− 18
Value of price change since 1973[a]	—	64	51	67	69	63	114	185
Current value	21	77	49	65	61	42	110	167
Services, interest & transfers	− 13	− 17	− 22	− 28	− 32	− 37	− 43	− 53
Current balance surplus	8	60	27	37	29	5	67	114
Export prices, index 1973 = 100	100	276	282	309	330	340	493	782
Import prices, index 1973 = 100	100	134	145	151	162	163	217	242
Terms of trade, index 1973 = 100	100	205	194	205	204	209	227	323

[a] Measured in terms of current quantities.
Source: OECD (July, 1980*b*, Table 62). Prices have been calculated residually.

After a small surplus of $8 billion in the OPEC balance of payments (current account) in 1973, the position improved drastically in 1974. The surplus gradually declined until 1978, when the order of magnitude of the $5 billion surplus was back at the 1973 level. The real increases in OPEC imports were 38 % from 1973 to 1974, and a further 38 % from 1974 to 1975. It is noteworthy that imports dropped 27 % in real terms from 1978 to 1979, after the sudden decline in the current balance surplus from 1977 to 1978.

Table 1 therefore suggests that the OPEC countries reduced their imports because of a budget constraint—the surplus on their current account balance of payments—indicating that the subsequent oil price increases in 1979 and 1980 could be viewed as an attempt to remove the budget constraint.

However, the figures in Table 1 are not representative of all OPEC member countries. The "high absorber" group[1] has exhibited a near-balance in their current account balance of payments during the period 1975–77, which de-

---

[1] The subdivision of OPEC member countries into a "high absorber" and a "low absorber" group is a classification used by the OECD Secretariat based on the development of imports after 1973/74; cf. Samuelson (1979). The "high absorbers" are Algeria, Ecuador, Gabon, Indonesia, Iran, Iraq, Nigeria and Venezuela. The "low absorbers" are Bahrein, Kuwait, the Libyan Arab Janahiriya, Oman, Qatar, Saudi Arabia and the United Arab Emirates.

teriorated to a large deficit in 1978, whereas the "low absorbers" have had large surpluses throughout. Thus, historically, OPEC imports have been less than their exports. But what factors have determined OPEC imports and what considerations have influenced OPEC price decisions?

Without attempting to answer these difficult questions in full, it will be argued briefly that OPEC decisions to raise the price of oil and to import can be viewed as parts of an overall strategy for economic development.

The OPEC organization is frequently referred to as a cartel which exercises monopoly power over oil prices. But this proposition may be questionable, although most observers would agree that the world market for crude oil contains strong oligopolistic elements.[1]

Unlike the monopoly firm, the final target of the OPEC countries may not be to maximize oil export revenues. Rather, oil revenues may be a means of maximizing national income before, during and after the transition from a resource-dependant to an industrial economy, if the consumption possibilities of the inhabitants are to be maximized over an infinite time horizon. Consequently, factors other than the marginal revenue and the marginal cost of producing oil enter into price decisions.

Many such factors should be allowed for simultaneously, e.g. the absorption capacity of the OPEC countries, the amount of foreign currency needed for domestic development projects, the growing elasticity of demand for oil as substitutes are developed and exploited, and the yield of monetary wealth versus the scarcity rent of oil resources.

It should be noted, however, that there are also vast differences among the OPEC countries as regards economic structure, the expected duration of oil reserves, oil resources per capita, etc. Hence, the optimal pricing paths differ among the OPEC countries and it is highly unlikely that any common price of oil exists which will maximize the needs of all countries simultaneously. On the other hand, large price differences between close substitutes, such as the various crudes, cannot exist on a permanent basis. Consequently, price determination within the OPEC organization will probably differ from the process suggested by the theory of imperfect competition.

The concept of respending is a convenient frame of reference for discussing OPEC imports during the 1970s. It is a one-word description of the idea that additional revenue from OPEC oil exports accompanying an increase in oil prices is followed by additional spending by the OPEC countries on imports of goods and services from the industrialized countries. The key elements in the definition of respending are the terms "additional" exports and imports.

---

[1] We will not attempt to analyze whether OPEC has been a market leader, thereby setting the price of oil, or has merely followed market trends. Judging from the suddenness of oil price developments during the 1970s, however, there is little doubt that the existence of OPEC is an important administrative element in the formation of oil prices, with an impact on the path, if not on the trend.

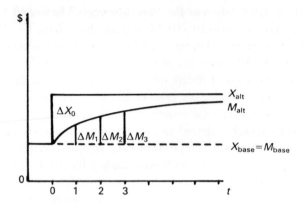

Fig. 1. The concept of respending.

Assuming an initial increase in OPEC export revenues of ΔX_0, the additional import value t periods later is ΔM_t. The increase in imports from the level that would have prevailed in the absence of an oil price increase, ΔM_t, can be calculated *ceteris paribus* as:

$$\Delta M_t = r_t \cdot \Delta X_0, \tag{1}$$

where ΔX_0: additional export value from the base; ΔM_t: additional import value in period t from the base; r_t: cumulated respending ratio in period t.

Assuming, initially, constant real exports and balanced trade the interpretation of the respending ratio is shown in Fig. 1. The cumulated respending ratio at period t can be estimated as:

$$\hat{r}_t = \frac{\Delta M_t}{\Delta X_0}. \tag{2}$$

If external balance is restored after n periods, then r_n will equal 100 %.

Empirical measurements of the "additionals" and thus of the respending ratio are not straight forward since they involve an evaluation of what the development would have been in the absence of oil price increases. Effectively, a measurement of respending involves a comparison of two different (at least one hypothetical) time paths of OPEC balance of payments, i.e. with and without an increase in the price of oil.

Another problem is the distinction between average and marginal respending, because it cannot be taken for granted that a given respending ratio is invariant to the magnitude of oil price increases.[1]

---

[1] The OECD has approached this problem by introducing "speed limits" on the real annual growth of OPEC imports, the effect of which is to reduce the respending ratio to additional export revenues, as the magnitude of oil price increases within a time interval grows; cf. OECD (July, 1980*b*).

Table 2. *Cumulated OPEC respending from 1974 to 1978, after the first oil price increase*

Cumulated figures, $ billions	1974	1975	1976	1977	1978
Additional exports, total OPEC	74	139	229	332	436
of which low absorbers	35	68	115	169	219
of which high absorbers	39	71	114	163	217
Addition current balance surplus, total OPEC	52	71	100	121	118
of which low absorbers	29	47	69	90	98
of which high absorbers	23	24	31	31	20
Additional imports, total OPEC	22	68	129	211	318
of which low absorbers	6	21	46	79	121
of which high absorbers	16	47	83	132	197
Respending ratio, total OPEC, %	30	49	56	64	73
of which low absorbers, %	17	31	40	47	55
of which high absorbers, %	41	66	73	81	91

Source: Table 1.
Note: Additional exports are calculated as the export in a given year less 1973 exports, after which these figures are cumulated.

Tables 2 and 3 contain very crude empirical estimates of the respending ratio, calculated in order to measure the possible differences in OPEC respending behavior after the first and second oil price waves.

The time paths compared are the actual development during the period 1973–78 and the hypothetical time path implying that the 1973 level of exports, imports and surplus on the current balance of payments would have prevailed in the absence of the oil price increases in 1973/74. On the basis of this heroic assumption, "additional" exports can be calculated as the OPEC exports in each year less the 1973 level of exports, by cumulating the figures over time. A similar procedure is applied to the OPEC current balance surpluses. The difference between cumulated additional exports and cumulated additional surpluses on the OPEC current balance of payments represents cumulated additional imports. Hence, the respending ratio—i.e. of cumulated imports to cumulated exports—will serve as our measure of respending in the subsequent discussion.

As can be seen from Table 2, the respending ratio was 30 % in 1974 after the huge price increases in the winter of 1973/74. Even this moderate respending implied an increase in the OPEC countries' import volumes far above what most observers had expected.

From 1974 to 1975, the cumulated respending ratio rose to almost 50 %. Since then, the respending ratio increased more slowly until 1978, when it was

Table 3. *Cumulated OPEC respending from 1979 to 1980 after the second oil price increase*

Cumulated figures, $ billions	1979	1980
Additional exports, total OPEC	66	225
of which low absorbers	38	124
of which high absorbers	28	101
Additional current balance surplus, total OPEC	63	172
of which low absorbers	27	84
of which high absorbers	36	88
Additional imports, total OPEC	3	53
of which low absorbers	11	40
of which high absorbers	− 8	13
Respending ratio, total OPEC, %	5	24
Low absorbers, %	29	32
High absorbers, %	− 29	13

Source and note: See Table 2 (1978 is the base year rather than 1973).

slightly less than 73 %. Average "low absorption" countries' respending was 55 % and that of "high absorption" countries over 90 %, according to Table 2.[1]

The evidence in Table 2 indicates that it has taken at least five years for OPEC to develop and implement domestic spending plans which absorb the steep rise in oil export revenues accompanying the first oil price "revolution".

The respending ratios, measured using the situation in 1973 as a basis, declined in 1979 and further in 1980 (not shown in Table 2). It therefore seems appropriate to regard 1978 as the final year in a cycle. Table 3 shows the respending for 1979/80 calculated on the assumption that the situation in 1978 would have prevailed in the absence of renewed steep price increases.[2] In all other respects the calculations are identical to those in Table 2.

In contrast to the development in 1973/74, respending in the first year after the new wave of oil price increases seems to have been much lower—5 % as compared to 30 % in 1974. The respending ratio for 1980 is expected to be 24 % as compared to 49 % in 1975.

Slower OPEC respending in 1979/80 cannot be attributed to the "low absorption" countries which respent about the same share of additional export receipts in 1980 as in 1975. A much more gradual response in the "high absorp-

---

[1] These respending estimates seem considerably lower than those employed by the OECD in marginal experiments, which for "low absorption" countries average a maximum of 70 % after one year, whereas the figure for "high absorbers" is 100 %; cf. OECD (July 1980*a*, p. 31).
[2] If the actual cycle exceeds five years, the use of 1978 as the new base year for calculating the respending ratios in 1979 and 1980 may bias the estimated ratio upwards.

tion" countries has caused slower overall OPEC respending. Part of the explanation may be the situation in 1979/80 in Iran—formerly one of the largest respenders.

There are many similarities between the first and second oil price rounds, also as regards the size of the external shock to the OECD countries. But the slowdown in respending constitutes an important difference that will cause the oil price impact to be felt differently this time, especially with respect to the relative impact on the USA and Europe.

III. OECD: Immediate Effects of an Oil Price Increase

The impact of an increase in the price of oil on the OECD economies can be roughly illustrated as the outcome of three elements which may vary in strength between countries. These elements are the terms-of-trade loss inflicted on net importers of oil, the OPEC demand for imports—respending—and the effect of the first two elements on the trading partners of the country or region in question.[1]

The geographical distribution of two of these elements is summarized in Table 4. The table contains figures for GDP (the most commonly used "weight" in aggregation), imports to OPEC from the OECD countries, and the terms-of-trade loss associated with a 10% rise in the price of oil over the average OECD oil import price in 1980, corresponding to an increase of slightly above $3.1 per barrel.

As can be seen from Table 4, there are considerable variations in the geographical distribution of GDP, OPEC imports and the hypothetical terms-of-trade loss. Although the US share of OECD GDP is 35.5%, its share of the terms-of-trade loss is smaller, 29.3%. The opposite is true for Europe, whose share of GDP is 42.6% and of the terms-of-trade loss, 47.8%. These figures reflect that European dependence on imported oil is larger than that of the USA, where indigenous production accounts for more than half of the energy requirement.

For net importers of oil, an increase in oil prices will imply a terms-of-trade loss and deterioration in the balance of payments. The terms-of-trade loss represents a transfer of income from the residents of each oil-importing country to the OPEC countries, as the distribution of world income changes.

The economic impact is much akin to the effect of an increase in taxation or a decrease in social transfers as regards the GDP effect (but not, of course, the balance of payments effect); GDP in real terms is not affected directly. Instead, the decline in GDP follows a reduction in real private spending that will take place as real private disposable income declines.

---

[1] Whereas the first two elements can be analyzed either directly or through econometric model simulations, world trade models have to be used to calculate the influence of the third element.

Table 4. *Geographical distribution of GDP, OPEC imports from the OECD and the direct terms-of-trade loss from a 10% oil price increase*

Percent of OECD total	Gross domestic product 1978	OPEC imports from OECD 1978	Terms-of-trade loss
USA	35.5	21.1	29.3
Canada	3.4	1.7	2.1
Japan	16.4	17.9	19.0
France	7.9	7.9	9.7
West Germany	10.7	15.5	11.9
Italy	4.4	8.9	6.6
UK	5.2	10.9	4.8[a]
Belgium–Luxembourg	1.7	2.8	1.4
Netherlands	2.2	3.1	1.9
Ireland	0.2	0.3	0.2
Denmark	0.9	0.7	1.4
Norway	0.7	0.2	− 0.7[a]
Sweden	1.5	1.4	2.1
Finland	0.6	0.4	1.0
Iceland	0.0	0.0	0.0
Austria	1.0	0.7	1.1
Switzerland	1.4	2.4	1.7
Spain	2.5	2.1	3.1
Portugal	0.3	0.0	0.3
Greece	0.5	0.5	0.5
Turkey	0.9	0.3	0.6
Australia	1.8	1.1	1.3
New Zealand	0.3	—	0.5
OECD			
Total	100.0	100.0	100.0
Europe	42.6	58.3	47.8
EEC	33.2	50.2	36.5

Note: The direct terms-of-trade loss is estimated as the immediate increase in currency expenditure on imported oil, assuming no change in oil imports in real terms.

The oil price increase is up 10% from the mid-1980 level. In the absence of 1980 data, the calculations have been carried out on a 1978-based model and data from 1978. Oil imports to the OECD countries in 1978 amounted to $139.1 billion (27.2 million barrels per day) or 2.3% of 1978 GDP for the total OECD. The oil import bill for 1980 can be estimated at $280 billion (24.3 million barrels per day) or 3.8% of 1980 GDP for the total OECD. It has been assumed that a 10% oil price increase from the mid-1980 level of $31.5 per barrel of crude oil corresponds roughly to a 16.5% oil price increase from the 1978 average oil import price level of $14.0 per barrel, when calculations are made using 1978-based data.

[a] Figures outdated due to large changes in domestic oil production since 1978.

OPEC respending increases real exports from the OECD countries and hence, directly, GDP at constant prices. In addition, real private spending will rise in response to the increase in export earnings, thereby stimulating real GDP indirectly as well.

According to Table 4, the immediate impact of an increase in OPEC respending should be more beneficial to Europe than to the USA due to the geographical distribution of OPEC imports from the OECD countries. The European countries deliver 58.3 % of all OECD exports to the OPEC area, as compared to the US share of 21.1 %, although their shares of OECD GDP are much more equal.

Thus, when two opposing forces are at work in the event of a rise in the price of oil, it cannot be concluded *a priori* that the effect of such an increase will be negative as regards real GDP in the oil-importing countries. This is provided, of course, that the individual countries are willing to accept, and the international capital markets are able to finance the corresponding balance-of-payments deficits.

In order to avoid a decline in real domestic demand, however, the marginal reactions of the industrial economies to external stimuli would have to be rather extreme. So for all practical purposes, it can be safely assumed that when oil prices are augmented, the standard of living as measured by domestic demand will decline in the industrial world.

Since the negative impact on real GDP of a $1 loss in the terms of trade is only indirect, whereas a $1 increase in real exports affects real GDP directly as well as indirectly, 100 % respending will most likely stimulate the real GDP in the industrial countries, as increases in exports more than offset decreases in domestic demand.[1]

The figures in Table 4 measure the immediate shock of oil price increases only. This kind of empirical analysis may not suffice to assess the impact of an oil price increase correctly as regards individual countries, since close international trade links may redistribute and alter the overall effect.

For instance, the initial terms-of-trade loss is partially offset through an increase in any one country's export prices, as the producers attempt to pass on the rising cost of oil. But since this stimulus to export prices will probably be present in all of the industrial countries, and since the OECD countries are knit tightly together through international trade, each country will experience a rise in the price of imported goods other than oil. The extent of this inflationary impact depends on the importance of oil to the trading partner economies.

---

[1] For purposes of illustration, this proposition may be regarded as analogous to the "balanced budget multiplier" theorem, provided, of course, that the parallel between the impact of real export changes and real government expenditure and between the impact of the terms-of-trade loss and taxes can be accepted with respect to real GDP. The analogy does not cover the balance-of-payments effect, however.

Similarly, the stimulation of increased imports to OPEC from OECD will spread from the countries which benefit directly from increased exports to their trading partners in the industrial world.

The use of models which incorporate international trade flows is one way of overcoming these difficulties.

IV. The Model System

The world trade model system (CRISIS) used to simulate the effects of a 10 % oil price rise is very similar to the OECD INTERLINK model.[1] The model treats the world economic system as a coherent and integrated whole; countries' trade flows and domestic economic trends are determined simultaneously. Basically, the model consists of:

— small structural models for each of the 23 OECD countries[2] which emulate the properties of larger national econometric models;
— reduced-form models for each of the 8 non-OECD regions, established on the basis of research carried out by the OECD and research centers specializing in the economies of non-OECD countries;
— a world trade and payments model which links the OECD economies and the non-OECD regions through merchandise and service trade volumes and prices.

Exchange rates, capital flows, factor payments and transfers are treated as exogenous.

The 23 OECD country models all display the same traditional Keynesian structure. Each consists of five import functions, a private consumption function, a private investment function, an income distribution relation and a price-wage spiral model which captures the effect of import price changes based on the mark-up principle. Export quantities and import prices are determined by partner countries' import quantities and export prices through market share matrices.

The determination of wages costs, and prices is one of the weakest parts of the model. The structural system is largely complete, but many of the parameters are based on inadequate evidence. The "sympathetic" response of the price of domestically produced energy to increased world energy prices is an area of particular uncertainty.

In contrast to the OECD INTERLINK model, the CRISIS model specifies government expenditures in real terms as exogenous. Consequently, an increase in the price level does not reduce real government spending. This difference

---

[1] Cf. OECD (July, 1980*b*) and Samuelson (1979).
[2] Belgium and Luxembourg are considered as one "country".

implies that the simulated effects of an oil price increase will be less depressive than the effects published by the OECD.[1]

The non-OECD region models are reduced-form "reaction functions". They determine import expenditure based on the regions' export revenues, i.e. the respending concept. Real imports will increase if real exports increase or due to terms-of-trade gains. The adjustments are assumed to run over several periods and are determined fundamentally by postulated respending lag parameters. But there are "speed limits" on the growth rates of imports, which enable the two OPEC regions in particular to accumulate huge reserves of unspent capital. The "speed limits" comprise the difference between marginal and average respending ratios.

These assumptions about respending lags and "speed limits" are key parameters with respect to the impact of an oil price increase on the OECD countries.

The world trade block that links countries' import quantities and export prices to export quantities and import prices through endogenous market shares is the greatest advantage of the model. Shares are functions of relative export prices and commodity composition. This construction allows for determination of the international impact on countries' terms of trade and domestic demand, including international repercussion effects. It is assumed that there exists only one world price for internationally traded oil.

The simplicity of the models for individual countries implies that the simulations described below should not be interpreted too rigidly as regards levels, but should be regarded as indicators of relative differences between countries.[2]

V. The Impact of OPEC Respending on the OECD

In order to assess the impact of OPEC respending on the OECD countries, experiments were carried out using the model system described above.

First, the impact of a 10 % increase in the price of oil from the 1980 average OECD import price is simulated under the assumption that OPEC respending follows the time path used by the OECD Secretariat for small oil price increases.[3] According to this assumed time profile, the respending ratio the first

---

[1] See Table 39 in OECD (July, 1980b, p. 129). There is no objective solution concerning the definition of the concept of unchanged economic policy. In some countries, it may be more appropriate to let nominal government expenditure remain unchanged instead of using the constant real terms approach applied in this article. The difference between adopting the constant nominal or the constant real terms approach, as regards the unchanged government expenditure assumption, may be an important factor in explaining the relatively small oil price effects calculated and discussed below, since the former assumption implies a reduction in government real spending in the case of an oil price increase in most industrial countries.

[2] The simulated overall impact of an oil price increase on the OECD countries suffers from at least one potentially serious omission from the model system, namely intra-non-OECD regional trade. Thus an increase in the price of oil does not affect these non-OECD economies to the extent that the region trades directly with OPEC.

[3] Cf. Table 31 in OECD (July, 1980a).

Table 5. *The simulated impact of a 10% oil price increase from the average oil import price level in 1980, with and without OPEC respending; first year effects*

	Change in real GDP, first year		
Percent	With 55% respending	Without respending	Effect of respending
USA	−0.06	−0.18	0.11
EEC	0.16	−0.14	0.30
OECD			
Europe	0.14	−0.14	0.28
Total	0.05	−0.16	0.21

year averages slightly less than 55%. The impact is calculated as the difference between a base line projection and a simulation where the price of oil has been changed *ceteris paribus*.

Then, the 10% oil price increase is again simulated; this time, however, OPEC respending is assumed to be zero.

The difference between the impact of an oil price increase with and without respending is calculated and can be interpreted as the impact of OPEC respending. The results discussed below refer solely to real GDP.

In the simulations, it was assumed that the individual OECD countries benefit from OPEC respending in proportion to their 1978 share of OPEC's import from OECD.

It should be noted that only the assumption about OPEC respending was altered. As mentioned in the preceding section, other regions of the world such as the Eastern European countries, Latin America, etc. also "respend" their export revenues. According to the model system, most of these regions improve their terms of trade *vis-à-vis* the OECD countries when oil prices go up. Therefore, the overall impact of an oil price increase differs from results that only take trade flows between OECD and the OPEC area into account.

Another important, although perhaps unrealistic, assumption concerns the politico-economic reaction to oil price increases and its effect on growth, inflation and balance of payments. Economic policy was assumed to remain unchanged, i.e. government expenditure on goods and services is not affected in real terms, taxes and duties are collected according to unchanged rates, and social transfers, e.g. to the unemployed, are paid out according to fixed rates.[1]

The simulation results are summarized in Table 5. According to the calculations, a 10% increase in the price of oil with 55% respending will affect real GDP in the OECD only moderately. The impact on the aggregate GDP may even be positive the first year, as both positive and negative GDP changes occur in the individual countries. Real GDP in the USA seems to decline

---

[1] Consequently, balance-of-payments deteriorations are not assumed to be neutralized by government policy.

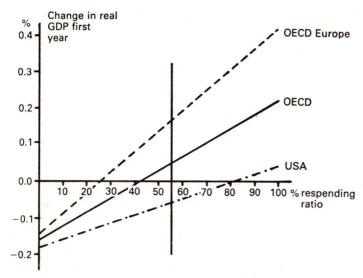

Fig. 2. The impact of a 10 % oil price increase under alternative respending assumptions.

slightly, as opposed to a somewhat larger European increase. In the absence of OPEC respending, almost all countries are affected negatively. The decline in real GDP in the USA of slightly less than 0.2 % is somewhat larger than the European decline.

The impact of respending is, of course, positive for all countries—the range of variation between countries is very wide. As could be expected from the distribution of OPEC imports (cf. Table 4), Europe is stimulated more, 0.3 %, than the USA, 0.1 %. Within Europe, the large EEC countries are affected the most.

The model simulations indicate quite clearly that the use of world trade models is important in evaluating the geographical distribution of the impact of respending. For large areas, such as a comparison between the USA and Europe, an examination of the OPEC countries' imports comes quite close to the simulated relative distribution of the respending impact. As regards the small European countries, however, their share of OPEC imports is only a poor indicator.

Large investments in world trade model construction and simulation therefore seem particularly justifiable for the small European countries, since this method significantly alters the assessment of the influence of worldwide phenomena on small open economies.[1] The simulation results in Table 5 can

---

[1] As such models are typically less sophisticated than large-scale national econometric models, direct use of their results for any individual country may be undesirable. In this case, the world trade model system can be used as an advanced export function, since it determines the imports of the trading partner countries. Then, the more accurate large national model can be run to calculate the national consequences of the change in exports suggested by the world trade model.

be illustrated as shown in Fig. 2 by interpolating and extrapolating from the two points determined by the simulations. Fig. 2 shows only linear approximations of the model simulations, but in this way the impact of a 10 % oil price increase for alternative responding assumptions can be readily and cheaply, although only roughly, assessed.

As is evident from Fig. 2, the impact line for Europe is steeper than for the USA; the impact line representing the OECD total is somewhere in between. Obviously, the original proposition that Europe will be affected more than the USA by a change in OPEC responding is confirmed by the simulations.

The figure also suggests the position of the very interesting intercept on the X-axis for each impact line. A responding ratio corresponding to the intercept of the impact line and the X-axis implies that real GDP in the first year is not affected by a 10 % oil price increase (provided that linearity is a reasonable approximation). As can be seen, this "neutral" responding ratio is as low as 25–30 % for Europe as compared to about 80 % for the USA, the point of "neutrality" for total OECD lies at a responding ratio of about 40–45 %.[1]

It should be emphasized that these conclusions cover first year effects only. In addition, if the figures had referred to domestic demand in real terms, the impact would have been negative, even in the case of 100 % responding.

VI. Conclusion

The overall impact of an oil price increase on real GDP in the industrial countries cannot be determined *a priori* because increased oil export revenues transferred to OPEC are in part returned to the industrial economies. The impact on GDP depends on whether this takes place in the form of responding, direct investment, portfolio investment or loans.

A more certain conclusion is that consumption in the industrial countries will be reduced. The stimulative responding part of the impact of an oil price increase is directed towards exports, thus increasing production but not (directly) domestic demand.

The sign and magnitude of the overall impact on GDP must be determined through empirical analysis. The discussion presented in this article serves to investigate the sensitivity of the GDP effects of an oil price increase to changes in OPEC responding. The analytical method of simulation using a world trade model improves the empirical determination of the impact—especially as regards small open economies.

An empirical measure of OPEC responding was constructed and computed for the two oil price hikes in the 1970s. The data for the period 1973–78 indicate that it took at least five years for responding to peak at 73 % for OPEC (91 %

---

[1] It goes without saying that this conclusion is subject to the credibility of the models used if any deductions about actual GDP changes are made.

for "high" and 55 % for "low absorbers"). Average respending over the five-year period was slightly less than 55 %. In the second round of oil price increase in 1979/80, the first year respending ratio was 5 %, as compared to 30 % in 1974. Similarly, the respending ratio for 1980 seems to be considerably lower than in 1975 — 24 % and 49 %, respectively.

A common notion is thus confirmed by more systematic analysis. However, it is noteworthy that the respending of the "low absorbers" was higher, 29 %, in 1979 as compared to 17 % in 1974. Lower overall OPEC respending may therefore be attributed more to less spending among the "high absorbers" in 1979/80 as compared to 1973/74.

In the numerous evaluations of world economic development in 1980, the oil-induced downturn is often compared to the 1973/74 situation. However, the present analysis points towards significant differences between the first and second oil price shocks.

The shock itself, i.e. the loss of real income in the industrial countries due to terms-of-trade losses, is about 2 % of OECD GDP in 1979/80, as was the case in 1973/74. But the overall GDP should have been more depressed in 1979/80 than in 1973/74 due to lower respending.[1] The downturn in Europe in particular should be deeper this time, because Europe is influenced more by the reduction in OPEC respending. As the large countries in Europe are hit the worst, this may be an additional explanation for the more evenly distributed balance-of payments deficits in 1979/80 than in 1973/74.

The assessment of respending in the 1970s and the simulation results (cf. Fig. 2) can be combined if the simulated first year effects can be regarded as representative of the medium-term impact—at least as regards the relative impact of the terms-of-trade loss and export increase.

According to the calculations, the impact of an oil price increase on OECD GDP turns positive, when the respending ratio increases above the 40–45 % range. Table 2 showed that the respending ratio passed the 50 % mark in 1975–76 and average respending during the period 1973–78 was about 55 %. Hence, real GDP in the OECD has not been reduced significantly, or has perhaps even been stimulated due to oil price development and respending since 1973—if the model assumptions and results are acceptable.

Nevertheless, the 3 % annual average real GDP growth in the OECD from 1973 to 1978 was lower than the $4\frac{3}{4}$ % per year experienced during the previous five-year period. It would be absurd to disregard oil price development as a major factor behind the slowdown in economic growth.

The OECD countries have been forced to adjust to the change in world income distribution by reducing real domestic demand and increasing the quantity of exports, while sometimes facing very large deficits in their balance

---

[1] That is, the development without any politico-economic reactions, since a change in policy response, as compared to events in 1973/74, may of course alter the observed economic development.

of payments and high inflation rates. The politico-economic responses to these hard facts have probably contributed significantly to the reduction in real economic growth during the readjustment period. The austerity measures introduced to defend national balance of payments and inflation goals are likely to represent the most important deviations from the model assumptions.

If the findings are reliable, the absence of concerted reflationary action on the part of the OECD countries, to ease the inevitable readjustment of the industrial countries from internal to external demand in response to higher oil prices, is even much more regrettable.

References

OECD: Fiscal policy simulations with the OECD International Linkage Model. *OECD Economic Outlook*, Occasional Studies. OECD, Paris, July 1980*a*.

OECD: *Economic Outlook*. OECD, Paris, July 1980*b*.

OECD: *National accounts of OECD countries 1950–1978*. OECD, Paris, 1980.

OECD: *Economic Surveys* (various countries). OECD, Paris, 1980.

Petersen, C. E.: CRISIS: An international transmission model. Institute of Economics, University of Copenhagen, Denmark 1980 (in Danish).

Samuelson, L. W.: The OECD International Linkage Model. *OECD Economic Outlook*, Occasional Studies. OECD, Paris, 1979.

LONG-TERM OIL SUBSTITUTION—
THE IEA-MARKAL MODEL AND
SOME SIMULATION RESULTS FOR SWEDEN*

Per-Anders Bergendahl and Clas Bergström

Energy Systems Research Group, Stockholm, Sweden

Abstract

The MARKAL model used in the International Energy Agency (IEA) Energy
Systems Analysis Project is surveyed in this paper. MARKAL is a multiperiod,
multi-objective linear programming model that represents a nation's energy
system. It includes new and conventional technologies, applies various policy and
physical constraints and optimizes in accordance with some specific criteria
(usually cost minimization). The cost-minimizing principle as a basis for a cooperative international strategy is also discussed. The trade-offs between system costs
and oil imports for Sweden are outlined. Some results concerning the structural
change in supply and demand technologies associated with these trade-offs are also
reported.

I. Background

In April 1976, the International Energy Agency established a Systems Analysis
Project. The primary objective was to evaluate the potential for new energy
technologies and thereby assist in formulating a strategy for reducing dependence on oil imports.[1]

A computer model, MARKAL, was developed which allowed an evaluation
of the potential impact of new technologies in competition with each other
and with existing technologies, under conditions related to various policy
and physical constraints. MARKAL is a multiperiod linear programming
model, capable of describing the evolution of widely diverse energy systems
under a variety of constraints and objective functions.[2]

---

* The authors are indebted to Lars Bergman, Alf Carling, Tord Eng, Karl-Göran Mäler,
Åsa Sohlman, Lewis Taylor, Göran Östblom and an anonymous referee for many useful
comments. Of course, the authors are responsible for any remaining errors.
[1] The participating countries and international agencies include: Australia (joined the
project in 1980), Austria, Belgium, Canada, the Commission of European Communities,
Denmark, Germany, Ireland, Italy, Japan, the Netherlands, New Zealand, Norway,
Spain, Sweden, Switzerland, the United Kingdom, and the United States. The project
was conducted jointly at Brookhaven National Laboratory (BNL) and Kernforschungs-
anlage-Jülich, Germany (KFA).
[2] By the end of 1979, fifteen countries had completed analyses using the MARKAL model.
The results for Sweden are summarized in the IEA Country Report for Sweden by Bergendahl (1980) and presented more extensively in Bergendahl & Bergström (1981).

The data which describe the various technologies were subjected to extensive, detailed review by the project staff and experts from industry and government institutions in the participating countries. A standard set of reference values which characterize the technology were adopted for many technologies. Individual countries have modified the reference values in order to account for variations in relative factor prices, geological conditions, specific national standards for environmental protection and public safety, etc. The reference and country-wise data are documented in Manthey (1979) and Sailor (1979).

II. The MARKAL Model

The energy system is a complex of interrelated markets that link alternative primary sources, processing, conversion and utilization technologies to a variety of end uses and user types. Since the feasible range of interfuel substitutability depends on the available supply and utilization technologies, the MARKAL model[1] is constructed around these technologies. A block diagram of the energy system as modelled by MARKAL is shown in Figure 1.

MARKAL includes a physical representation of technologies, energy flows and conversion efficiencies in various stages from extraction or importation to the ultimate goal of satisfying a given demand for useful energy. Transformation technologies convert primary energy carriers into intermediate energy forms, which are in turn transformed through different demand technologies into useful energy such as mechanical energy and space heating.

The Swedish version of MARKAL includes 23 groups of new technologies (see Appendix). The groups are intended to cover the major new technologies. Each technology group may contain one or more specific processes. For example the first group, coal liquefaction, contains four components: hard coal methanol production, peat methanol production, hard coal hydrogenation and hard coal liquefaction (Fisher–Tropsch). Each of these is described by data set which includes: technical features (efficiency, availability factor, technical lifetime, type of fuel inputs and outputs); cost data; and constraints on installed capacity.

Fig. 1 is basically an aggregated flowchart that traces "all" of the possible paths by which a joule of energy can find its way through the system to satisfy useful energy demand.

Each path, from a primary energy source to useful energy, has corresponding supply efficiencies and a demand efficiency. Supply efficiency is the proportion of primary energy not lost when it is converted into an intermediate

---

[1] Many features of the model were adopted from existing models at the two cooperating laboratories: BESOM and DESOM, the two Brookhaven energy systems optimizing models and the KFA Energy Supply Optimization Model. Additional modelling was undertaken in order to meet the specifications of the Energy Systems Analysis Project.

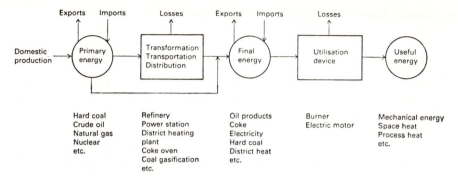

Fig. 1. Block diagram of the energy system as modelled by MARKAL.

energy form. Demand efficiency is the energy not lost as it is converted into useful energy. The set of all paths, from the sources to the demands, forms a reference energy system (RES).

The MARKAL model is designed around the RES. The major types of constraints in MARKAL are:[1]

Energy Carrier Balance Equations: This group of equations ensures that imports and production for each energy carrier and each time period are greater than or equal to total consumption (i.e. consumption by utilization devices, input to processes and conversion systems).

Electricity Balance: Electric production from all generating plants is equal to or greater than the total demand for electricity. There is one equation for each year and time division.

Balance for District Heat: Production of low-temperature heat is greater than or equal to the consumption of low-temperature heat in each time division and time period.

Base-Load Equations: Night electricity production from base-loaded electric conversion technologies has to be less than or equal to a fraction of the day electricity production. The reason for this equation is that base-loaded production must be constant between day and night because these technologies cannot follow the load.

Capacity Transfer Equation: These inter-period constraints relate activities across time periods. They ensure that the residual capacity (initial capital stock corrected for depreciation) plus the investment made for every technology in the current and previous time periods is greater than or equal to the installed capacity of the technology in a given time period.

Electric Peaking Equations: Installed electric generating capacity must be adequate to meet the peak, in the season and time of the day, of the greatest load.

---

[1] See also Abilock & Fishbone (1979).

Salvage Equations: This group of equations provides for recapture of the remaining life of the capacities. These equations are basically a correction that reduces the cost for investment in a technology, when part of its technical lifetime will extend beyond the model horizon.

Demand Equations: For each demand sector, these equations ensure that the end-use energy output from each of the demand devices is greater than or equal to the end-use demand.

World market prices of energy carriers, technical characteristics, cost data and demands for useful energy are exogenously given components of the model. However, exogenously given demand for useful energy does not mean that energy demands at earlier stages are unresponsive to prices. The model endogenously determines the demands for gasoline, electricity, light distillates, etc. while the demands for energy by end-use category, such as travelling or space heating, are exogenously specified. This allows for interfuel and capital-energy substitution in particular end uses.

There are two categories of exogenously given costs, annual and investment. Annual costs include mining, imports, operation and maintenance (fixed and variable), as well as transmission, distribution and delivery. Investment costs cover all technologies including transmission and distribution grids for conversion technologies. Capital stock in the initial year is regarded as a sunk cost.

Given the set of constraints, there are of course many technically feasible ways of meeting the exogenously given demand for useful energy. MARKAL is a multiperiod linear programming model whose "purpose" is to select an optimal pattern of energy flows in the RES, in accordance with some specific criteria.[1]

A principal optimization criterion is, obviously, cost minimization, where the objective is to minimize total discounted system costs. Accordingly, a solution is given whereby the cost-minimizing paths and associated technologies and primary energy sources are selected.[2] A set of endogenously determined dual variables complements such a solution. These variables are the efficient prices imputed to energy carriers, demand categories, etc. by the economic (cost-minimizing) criterion governing the allocation of scarce resources under given technical and policy constraints.

Thus, although the model focuses on the technical structure of the energy system, the solution (according to the dual formulation) depends on the economic criterion that no activity should be used if a more profitable activity or combination of activities is available to the nation.

---

[1] The Swedish nine-period (45-year) model has approximately 3 600 constraints and 4 200 variables.

[2] Of course, the choice of discount factor is of strategic importance as it implies a specific weighting between investment costs and annual costs. High discount rates increase the weight of investment costs in relation to annual costs. Hence, given cost minimization, a higher discount factor will discriminate against the introduction of capital-intensive technologies and to some extent "freeze" the prevailing structure; cf. Section IV.

III. Multi-objective Analysis and International Aggregation

Initially we assume that planning of the future energy system is based on multiple criteria. Given a specified set of such criteria, we seek a procedure for estimating the potential conflicts between these objectives as well as their corresponding quantifications. In other words, we want to derive a scheme which displays the trade-offs between the various objectives as a basis for policy evaluation. What needs to be defined then, is the feasible area in an n-dimensional plane, where n is the number of specified criteria.

It should also be emphasized that this analysis only defines the feasible range of efficient trade-offs between various objectives and is not concerned with the problem of selecting an optimal point in n-criteria space.

Let us now assume that society is willing to pay a premium for reducing oil imports. We limit the problem to include only two objectives, costs and oil imports. In this way, we also gain the advantage of a simple graphic representation. To be more specific: assume that accumulated oil imports between periods 1 and T comprise one criterion and that the total discounted system cost is the other criterion. The MARKAL model can now be used to derive the efficient trade-off locus for oil imports and costs.

The solution to the cost-minimization problem determines a certain (total discounted) cost, which we denote π^*, as well as a quantity of (accumulated) oil imports, x_k^*. If the objective function is changed to represent a minimization of oil imports, we get a solution which involves a minimum amount of oil imports (x_k^{**}) and a corresponding cost figure π^{**}.[1]

We are now equipped with the extreme points on the efficient part of the feasibility locus. The intermediate points on the locus can be derived in any of the following three ways: (i) successive changes in a constraint on oil imports in the $x_k^* - x_k^{**}$ segment and cost minimization; (ii) successive changes in a constraint on costs in the $\pi^* - \pi^{**}$ segment and oil import minimization; and (iii) successive changes in the relation between α and β, given the objective function Min $\alpha\pi + \beta x_\kappa$.

When only a limited number of points on the curve are derived, the choice of approach is relevant for aggregation purposes. The third approach was chosen here, as it makes structural information available at points where the marginal cost of oil import reductions is the same for all countries. Such a criterion for aggregation (at equal marginal cost) implies a rational strategy for the group of countries with respect to cost-minimizing cooperative efforts to reduce their total oil imports.

Assume that two countries, A and B, exhibit quite different curvatures of their trade-offs between costs (π) and oil imports (x_k).[2] This means that the

---

[1] For purposes of illustration, we assume that this problem gives us a unique solution.
[2] This discussion is based on Bergendahl & Teichmann (1980).

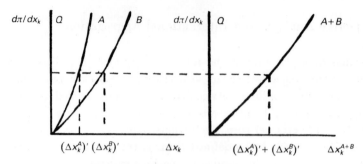

Fig. 2. Individual and aggregate marginal costs.

marginal cost $(d\pi/dx_k)$ structure for reducing oil imports from the minimum system cost level differs in the two countries. This is schematically illustrated in Fig. 2, where Δx_k symbolizes $(x_k^* - x_k)$.

Each point on the aggregate marginal cost curve represents the cost-minimizing reduction in oil imports for the group of countries as a whole. The underlying implementation scheme for technologies implies disproportionate involvement for the different participants. Those participants with the greatest oil import reductions at a given marginal cost have the largest share in the technology implementation plan and vice versa.

If we assume the existence of a perfect oil market (no extraneous incentives to decrease oil imports, correct anticipation of future oil prices, etc.), the minimum cost solutions should of course be chosen by each participant. Let us assume, however, that a premium is attached to oil import reductions as a reflection of market failures.

Without cooperation, each country would be responsible for adjusting its own level of imports in accordance with its own risk aversion and evaluation of market failures. However, given cooperation, a joint objective is assumed to be set up with regard to the reduction in aggregated oil imports. Whatever the magnitude of this reduction, each point on the aggregate trade-off curve implies a certain international allocation of oil import reductions via a specific plan for implementation of technologies. Normally this efficient plan differs from the "proportional reduction" strategy, which is today's guiding principle in practical policy design (cf. Section IV).

If there exists a joint strategy for oil import reduction, the oil price assumptions in MARKAL become somewhat dubious. Oil prices are assumed to change over time in a way that is not influenced by the IEA countries' choice of energy sources.[1] So long as the MARKAL approach is applied to the policies of a single, small country such as Sweden, this is quite reasonable. But it does

---

[1] This of course does not preclude sensitivity analyses where the effects of different (exogenous) oil price scenarios are studied.

not seem realistic to assume that a powerful import-reduction policy by the IEA group as a whole (or by its larger member countries) would not influence OPEC's price policy for crude oil.

However, an alternative treatment of oil prices cannot be based on simple monopoly or monopsony models—combined with assumptions of changes in the relevant elasticities between the different cases. The market situation would essentially be characterized by bilateral monopoly, complicated by potential conflicts within each of the two groups (OPEC and IEA). A rather advanced, game-theoretical approach would be required for a realistic treatment of the market. This approach, in turn, would be very difficult to combine with a large-scale, linear model such as MARKAL.

We do not discuss oil market problems any further, but concentrate on the choice of energy supply strategy for an individual country (Sweden), where the oil price assumptions do not present problems of this kind.

IV. System Costs and Oil Imports

The trade-off between system costs and oil imports for Sweden is illustrated in Fig. 3. The cost and import figures are totals for the period 1980–2020. When the objective is pure cost minimization, the total discounted system costs are \$133 billion and accumulated oil imports are 32.4 EJ.[1] Emphasizing the other extreme objective, i.e. oil import minimization, the system gets along with total oil imports of only 22.7 EJ. This occurs at the expense of increased total system costs, which now amount to \$144 billion. The intermediate case \$1/GJ is derived by setting $\alpha = \beta = 1$ in the objective function $\alpha\pi + \beta X_k$ (see Section III).[2]

Fig. 3 clearly illustrates an increasing marginal system cost associated with making the energy system less dependent on oil imports. For example, with the cost-minimization scenario as a point of reference, a first reduction in total oil imports by about 5 000 PJ means an increase in the total system cost of less than \$1 billion. But an additional reduction of the same magnitude brings about an increased system cost of as much as \$10 billion. The tech-

---

[1] (i) The standard discount factor was 6 percent. Taking the cost-minimizing case as a point of reference, sensitivity studies were made using a lower (3 %) and a higher (10 %) discount factor. It was found that the share of investment costs in total system costs decreases from 26 percent in the 3 % case to no more than 17 percent in the case with the highest discount rate. Total accumulated oil imports are 11 percent lower in the 3 % case as compared to the 10 % case.

(ii) The "cost min" case involves the assumption that real oil prices will rise throughout the time span by about 2.6 percent each year. Movement along the trade-off curve away from the "cost min" case represents a gradually increasing surcharge on these oil prices.

(iii) All prices are in 1975 US dollars unless otherwise noted. EJ means exajoules (10^{18} joules) and accordingly, GJ denotes gigajoules (10^9 joules) and PJ petajoules (10^{15} joules).
[2] The curve is drawn on the basis of two additional points, \$0.5/GJ and \$1.5/GJ, respectively, i.e. $\alpha = 1$ and $\beta = 0.5$ and 1.5, respectively. Thus, strict convexity is an assumption.

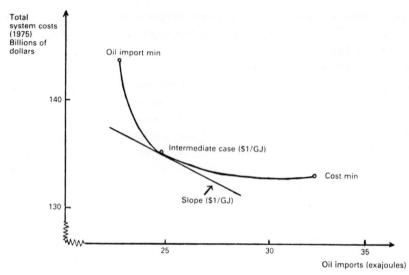

Fig. 3. Trade-off curve between costs and oil imports.

nological "mirror image" of the (flat) portion of the trade-off curve associated with relatively low marginal costs is, of course, the fact that some technologies prove competitive at very moderate additional increases in oil prices. Some changes in the technological structure along the trade-off curve are discussed in Section V.

In order to compare our own situation with that of another oil-dependent nation and the IEA group as a whole, and to provide empirical evidence for the discussion at the end of Section III, we have calculated normalized system cost and oil import indices.

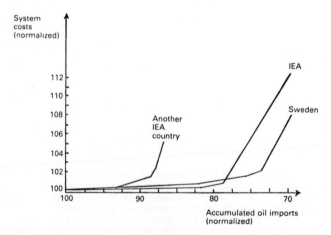

Fig. 4. Normalized scenario indicators.

Table 1. *Oil imports: 1980 in the "cost min" case is normalized to 100*

Year	Base case cost min.	Intermediate case ($1/GJ)	Minimizing oil imports
1980	100	99	95
1990	82	66	56
2000	60	49	45
2010	58	33	31
2020	43	18	18

In these normalized scenario indices, the results for Sweden and the IEA countries are referred to the cost minimum case with index 100 for both system costs and cumulative oil imports. (Fig. 4). The maximum reduction in oil imports varies between the IEA countries. The aggregate IEA analysis indicates that cumulative oil imports can be reduced by about 30 percent for the IEA as a whole by incurring approximately a 12 percent increase in energy system costs. The maximum reduction in oil imports for the other IEA country is only 14 percent at a cost increase of 5.6 percent. The figures for Sweden are 30 and 8 percent, respectively.

Furthermore, the slope of the trade-off curve also varies between countries. Thus, if the oil-consuming countries have agreed to reduce their total oil imports, a proportional distribution of oil reductions would not be an optimal strategy with respect to costs.

Table 2. *Primary energy mix—1990, 2000 and 2010, along the trade-off curve*

	1990			2000			2010		
	Base case, cost min	Inter- mediate case ($1/GJ)	Mini- mizing oil imports	Base case, cost min	Inter- mediate case ($1/GJ)	Mini- mizing oil imports	Base case, cost min	Inter mediate case ($1/GJ)	Mini mizing oil imports
Coal	8	12	15	19	23	23	18	31	30
Oil	39	31	26	27	21	19	23	12	11
Gas	1	2	2	2	1	1	2	0	0
Nuclear (FEQ)[a]	16	19	20	16	19	19	21	24	23
Hydro- electric ((FEQ)[a]	25	24	25	24	23	24	22	20	21
Renewables	11	11	11	13	12	14	15	14	15
Total primary energy (FEQ)[a]	100	100	100	100	100	100	100	100	100

[a] Fossil fuel equivalents.

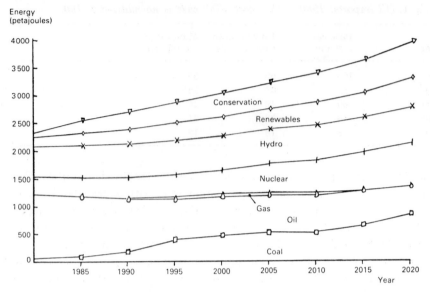

Fig. 5. Primary energy supply, FEQ[1] (PJ/YEAR), in the cost minimum case.

In Table 1, Swedish oil imports in 1980 in the cost minimization scenario are normalized to 100. The switch away from oil as the basic energy carrier is shown to be a feature of the "cost min" case in itself, whereby imports in the last period are only 43 percent of the initial amount. When the objective is minimization of total oil imports, they declined to not more than 18 percent of the 1980 level.

As shown in Table 2, interfuel substitution significantly affects the optimal energy mix in the years 1990, 2000 and 2010. For reference, the primary energy supply in the cost minimum case is illustrated in Fig. 5. Of course, the most important substitution takes place in the last year reported here. The most striking shifts are steady decreases in the share of oil usage and steady increases in the shares of nuclear power and coal. Natural gas does not seem to be competitive, regardless of the premium paid for oil import reductions.

Oil usage decreases from 39 percent of total primary energy in 1990 to 23 percent in 2010 in the scenario that emphasizes cost. The corresponding figures for the minimum oil import scenario are 26 and 11, respectively.

Coal usage increases as we move along the trade-off curve from 8 percent to 15 percent in 1990. The figures for 2000 are 19 and 23, respectively. In 2010 the share of nuclear power is 24 percent at the north-west extreme of the trade-off curve. The share for nuclear power in 1990 in the cost minimum case is only 16 percent.

Interfuel substitution dampens the initial "shock" as it is transmitted through the system. The shock is brought about by changes in the structure

Table 3. *Shadow prices of fuels in the intermediate case ($1/GJ) (normalized to the corresponding values in the cost minimum case)*

	1990	2000	2010
Methanol	110	127	117
Light distillate	174	188	168
Gasoline	143	151	168
Electricity	108	100	104

of secondary energy production and utilization technologies. These shifts are, of course, attributed to relative fuel price development as shown in Table 3 for a sample of fuels included in the model.

V. Transformation and Utilization Technologies

Interfuel substitution results from changes in the structure of transformation and utilization technologies. We now turn to the change in the mix of some supply and demand technologies between the end points on the trade-off curve in Fig. 3.[1]

V.1. *Electricity Production*

A comparison of the "cost min" and "oil import min" scenarios indicates that the structure of electricity production is rather insensitive to cost or oil import optimization. In both cases hydroelectric power and nuclear power are the main contributors. Furthermore, the following technologies penetrate the market rather early and remain important: a) industrial back-pressure power stations based on biomass; b) coupled electricity and heat production based on hard coal; and c) coupled production based on peat.

However, when the oil import objective is stressed, LWR coupled production with pass-out turbines is introduced on a larger scale at an earlier stage.[2]

---

[1] For a complete display of technological change along the trade-off curve, see P. A. Bergendahl & C. Bergström (1981).

[2] A "constrained nuclear" scenario illustrated the evolution of the electricity system when extensive use of both oil and nuclear power is prohibited. This feature is invoked via a limit on the use of nuclear power in the intermediate case ($1/GJ). The constrained use of nuclear power is modelled as an upper bound on the cumulative use of nuclear power. The upper bound is taken to be 65 % of the use in the intermediate case. Also, as could be expected, this scenario is based on hydroelectric power. The limit on nuclear power is not stringent enough to prevent the LWR technologies from becoming vital to the system during the next 25 years, but after that these technologies are only of minor importance. Furthermore, the fast breeder reactor is not introduced into the system as in the "cost min" and "oil import min" scenarios.

The mirror image of a heavy decrease in the use of nuclear power at the beginning of the next century is greater diversification of the electricity system. Coal is thereby used more intensively. The following coal technologies are particularly important: coal MHD (magnetohydrodynamic) electric production; coupled production based on hard coal; and a hard coal combined cycle.

The wind electric power plant also becomes an important technology during the latter part of the 45-year period.

This case also involves utilization of a higher cost hydroelectric technology and an earlier reduction in oil-based electricity production.

V.2. *Space Heating*

The most obvious structural changes concerning the technologies which contribute to meeting the residential demand for space and water heating in the cost-minimization case are:

(i) a decline in the use of oil burners to about 5 % of the market in the year 2000, as compared to about 60 % in the year 1980;

(ii) a gradual increase in the use of district heat exchangers from a share of 20 % in 1980 to about 50 % in 2000 and thereafter;

(iii) considerable efforts to increase energy efficiency via insulation, heat management, etc.; from 1990 the effects of these measures correspond to 20–30 % of the energy consumption for space and water heating (according to previous standards);

(iv) a gradual increase in the use of solar technologies from around year 2000, to reach a market share of around 10 % in 2020; and

(v) an increase in the use of gas burners and electric heat pumps and a decrease in the use of electric resistance technologies.

When the objective is switched to oil import reduction, the induced structural changes are similar to the "cost min" case, but intensified. Thus, oil burners are taken out of use earlier and district heat exchangers are used more intensively during the 1990s. At the end of the 45-year period, there is also greater use of gas burners and solar technologies. The contribution from solar technologies is approximately 16 % around the year 2020, all of which is attributed to use in new single-family houses.

V.3. *Road Transportation*

The energy demand for road transportation is at present totally met by oil products. When oil imports are reduced, fuel demand grows more diversified, however, via switches in utilization technologies.

In the cost min case, there is a continuous demand for gasoline and light distillate (diesel) throughout the period. However, the use of light distillate takes an increasing share of the market. Methanol does not become an important fuel until around the year 2020, but by then it accounts for almost 40 % of the market.

The overall picture reflects uninterrupted use of conventional vechicles, i.e., diesel and Otto engines. The demand for methanol at the end of the 45-year period is due to the use of the Otto engine fueled with pure methanol. The mix of gasoline and methanol has not proved efficient in this scenario.

In the trucking market, the Stirling engine (fueled with light distillate) is introduced on a rather small scale around the year 2000 and subsequently used more extensively to account for slightly over 10 percent of the truck engines in 2020.

When the objective is switched to oil import minimization, a more diversified fuel demand evolves owing to an increasing use of non-conventional technologies. In this case, the use of gasoline declines rather rapidly to a market share of roughly 10 percent in the year 2010. The use of light distillate also diminishes after 1990. Instead, the use of methanol accelerates from around 25 percent of the market in the year 2000 to as high as 74 percent in 2020. A demand for electricity and hydrogen is also part of the considerably changing market structure.

Concerning the truck fleet, use of the conventional "diesel/Otto engine" diminishes and is gone from the market by around 2010. There is instead an expanding market for the diesel engine with double injection. The Stirling engine, which penetrates the market to some extent in the cost minimum case, does not become a dominant technology even in this scenario, where it reaches its highest market share in 2010 with about 30 percent of the truck market.

With regard to private cars, there is an extensive use of the gasoline–methanol mix as early as the 1990s. However, the use of this mix is principally a transition phenomena in the process of replacing conventional diesel and gasoline engines with the Otto engine fueled by pure alcohol. This technology takes a dominant position around year 2020.

By 2005, the battery car claims a considerable portion of the car fleet and accounts for almost 30 percent of the market (as measured by energy demand) in 2020. The "electrical fuel cell" is used to a limited extent at the end of the 45-year period.

VI. Useful Energy Demand

The MARKAL model treats interfuel substitution in detail and the cross elasticities which arise from changes in relative fuel prices are implicitly given in the optimization. However, the model lacks a representation of the interplay with the rest of the economy. Thus, the impact on the level and composition of real GNP is not caught. The demand for useful energy is treated as an exogenous datum; changes in relative prices do not induce any changes in this demand vector.

Table 4 gives an idea of this incomplete specification of the quantitative relationships between the energy sector and the rest of the economy. It displays the changes in marginal cost from satisfying useful energy in some of the demand categories modelled in MARKAL. Note in particular the marked long-run increases in shadow prices for "transport energy" when oil imports are limited. By assumption, changes in relative prices do not induce any re-

Table 4. *Changes in the marginal cost from satisfying useful energy demand in the intermediate case (percent of the corresponding values in the cost min solution)*

Demand Category	1980	2000
Industry	+39	+4
New single-family houses	+8	−3
New multi-family houses	+20	−11
Old single-family houses	+24	−5
Old multi-family houses	+27	−12
Commercial	+26	−4
Service	+27	+6
Ship transport	+31	+69
Rail	+28	+8
Auto	+23	+44
Truck	+26	+23
Bus	+35	+63
Air transport	+26	+47

allocation of personal transportation (between cars, rail, bus and air) or of goods transportation (between truck and rail).

As a result, the ways in which alternative energy policies and external changes affect the economy cannot be fully described. On the other hand, economic growth models have not in general considered the energy sector in sufficient detail.

An important research topic may be to analyze the specification of the quantitative relationship between the energy sector and the economy in general. Special reference could be addressed to a possible iterative combination of the Swedish version of MARKAL with the quantitative general equilibrium model of the Swedish economy.[1] Such a combination might be able to provide a comprehensive long-run representation of the relationship between the energy sector and the rest of the economy.[2] But as a word of caution, it should not be taken for granted that one giant model is the best tool for analyzing impacts at different levels of aggregation.

Appendix

Coal Technologies

1. Coal liquefaction
2. Coal gasification

---

[1] See Bergman (1980).
[2] Such a framework for assessing the full range of alternative energy policies has been created by Hoffman & Jorgenson (1977) using an approach that encompasses the DESOM (10) model—MARKAL's predecessor—and the econometric model of interindustry transaction developed by Hudson and Jorgensson (1974).

3. Coal combustion (fluidized bed)
4. Magnetohydrodynamics (MHD)
5. Fuel cell

Solar technologies

6. Residential and commercial solar
7. Wind power
8. Ocean power
9. Fuels from biomass
10. Dispersed solar electric

Other

11. Non-fossil hydrogen systems

Nuclear Technologies

12. Advanced converter reactors
13. Breeder reactors
14. Fusion

Geothermal

15. Hot dry rock

Conservation residential and commercial:

16. Building efficiency (shell)
17. Building efficiency (equipment)

Transportation

18. Improved efficiency
19. Alternative fuels
20. New systems—electric auto
21. New systems—other

Industry

22. Heat management
23. Process specific

References

Abilock, H. et al.: A multiperiod linear programming model for energy systems analysis. BNL 26390, Oct. 1979.

Abilock, H. & Fishbone, L.: Users' guide for MARKAL (BNL-version). BNL 27075, Dec. 31, 1979.

Bergendahl, P. A.: Energy systems analysis for Sweden 1980–2000. A multiobjective analysis of options, contraints and the contribution from new technologies. BNL 27620, Feb. 1980.

Bergendahl, P. A. & Bergström, C.: *Longterm energy options for Sweden.* The IEA model and some simulation results. DFE Report No. 36, Stockholm, 1981.

Bergendahl, P. A. & Teichmann, T.: A cost-minimizing strategy to reduce oil imports: paper presented at the IAASTAD Conference in Montreal, May 1980.

Bergman, L.: Energy policy in a small open economy: The case of Sweden. IIASA Report 1, No. 1 (January–March 1980), 1–48.

Hoffman, K. C. & Jorgenson, D. W.: Economic and technological models for evaluation of energy policy. *Bell Journal of Economics 8*, No. 2 (Autumn 1977), 444–466.

Hudson, E. A. & Jorgenson, D. W.: U.S. energy policy and economic growth 1975–2000. *Bell Journal of Economics and Management Science 5*, No. 2 (Autumn 1974), 461–514.

Manthey, C.: *Technology data handbook*, Vols. 1 & 2. Jülich, Dec. 31, 1979.

Marcuse, W., Bodin, L., Cherniansky, E. A. & Sanborn, Y. A.: Dynamic time dependent model for analysis of alternative energy policies. BNL 19406, 1975.

Sailor, V. L.: Technology review-report (revised). BNL-27074, December 1979.

THE EFFICIENCY-FLEXIBILITY TRADE-OFF AND THE COST OF UNEXPECTED OIL PRICE INCREASES*

Lars Bergman and Karl-Göran Mäler

Stockholm School of Economics, Stockholm, Sweden

Abstract

The starting point for this article is the notion of a possible trade-off between static efficiency and *ex post* flexibility of input proportions. It is found that under reasonable assumptions, plants designed for a single input price constellation can be made more efficient than those designed for a variety of input price constellations. In order to obtain an estimate of the economic significance of flexible input proportions *ex post*, a simulation model of the general equilibrium type is used to analyze the impact of unexpected oil price increases. The results indicate that, compared to a case with rigid technology, the social cost of an oil price increase is significantly lower when the technology is flexible. However, other rigidities, e.g. on the labor market, seem to be more significant than technological rigidities in this context.

I. Introduction

There seems to be common agreement that one of the major problems in Swedish energy policy is the economic risk stemming from the country's high dependence on imported oil in conjunction with uncertainty about future oil prices. The policy measures taken so far in order to reduce this risk are all aimed at reducing Sweden's consumption of oil. More specifically, the introduction of oil-efficient technologies has been encouraged, and additional policies along the same lines are planned for the future.

In this study we approach the problem of reducing the economic risk associated with Sweden's dependance on imported oil from a slightly different point of view. Starting from the notion that the properties of technology at the *ex ante* stage differ from the properties *ex post*, we consider a possible trade-off between static efficiency and *ex post* flexibility. The idea is that plants specially designed for a single input price constellation can be made more efficient than those designed for a variety of input price constellations.

---

* Financial support from the Energy Research and Development Commission is gratefully acknowledged. We would also like to express our gratitude to Andras Por at IIASA, Laxenburg, Austria, for development of the solution algorithm which is an essential part of the simulation model used in this study.

From the energy policy point of view, this means that there is a trade-off between at least two types of policies aimed at reducing the economic consequences of unexpected oil price increases. One type, obviously, is to stimulate the use of oil-efficient technologies. But these technologies might be quite rigid in terms of oil input coefficients.

The second type of policy, then, is to stimulate the use of technologies where oil input coefficients are also relatively flexible *ex post*. If oil prices attain their expected values, the second type of policy would entail higher oil consumption and lower real income. This is because *ex post* flexibility can be attained only at the expense of static efficiency. However, if oil prices reach unexpectedly high values, a technology with relatively high *ex post* flexibility would make it possible to reduce the use of oil and thus mitigate the real income loss due to the higher oil prices. Obviously, the optimum mix of conservation and flexibility-increasing policies depends on the probability distribution of future oil price developments and the cost of achieving various degrees of *ex post* flexibility in oil input coefficients.

The purpose of this article is to discuss the efficiency-flexibility trade-off in a formal setting and to present some estimates, based on Swedish data, of the social benefits of a flexible technology when oil prices attain an unexpectedly high level. Since there are no acceptable estimates regarding the cost of various degrees of *ex post* technological flexibility, our results should be regarded primarily as illustrations of a principle; consequently they do not lead to definite policy conclusions. As our analysis essentially concerns the economic significance of technological rigidities, we have tried to place the numerical results in perspective by comparing them with estimates, derived in the same way, of the impact of labor market rigidities.

II. Efficiency versus Flexibility

We begin by discussing the problem of efficiency versus flexibility from a theoretical point of view. So as not to overburden the presentation, we consider only one sector and one energy input, E, with uncertain price Q. In addition to E, labor, L, is the only variable input with price W, assumed to be known with certainty. The third input is capital, K, where the user cost is equal to the real interest rate, R,[1] also assumed to be known with certainty. We assume that the amount of capital cannot be changed after the plant has been built. K should be interpreted as an aggregate of different capital goods; there are in general several different ways of combining these goods to yield the same K. A design variable, D, is assumed to reflect this diversity of K. It is assumed that D has to be determined prior to the actual construction of the plant. The output from the plant is denoted X.

---

[1] Thus, for simplicity, we disregard depreciation and the possibility of capital goods prices differing from the price of output in the sector under study.

The plant is completely characterized by the vector $Y = (X, Z, D)$, where $Z = (L, E, K)$ is the input vector. In order to be technically feasible, Y must belong to the feasibility set T, that is $Y \in T$. It is assumed that T is a convex, closed cone, i.e. in the long run there are constant returns to scale.

The long-run production function is given by

$$F(L, E, K) = \sup \{X; \exists D, \text{ such that } (X, L, E, K, D) \in T\}. \tag{1}$$

As T is a cone, it follows that F is linearly homogeneous in Z. Since T is closed, the supremum will be attained. The set of D that maximizes (1) is denoted $D(Z)$. We can now define the production set $\overline{T}$ as the technically efficient set

$$\overline{T} = \{(X, L, E, K); (X, L, E, K, D(Z)) \in T\}. \tag{2}$$

When complete adjustment is possible, the cost function is given by

$$TC(X, W, Q, R) = \inf \{WL + QE + RK; (X, L, E, K, D) \in T\}. \tag{3}$$

As T is closed, the infimum is attained. Since T is a cone, it follows that TC is linear in X, so that the unit cost can be defined as

$$(1/X) TC(X, W, Q, R) = \varkappa(W, Q, R), \tag{4}$$

which is independent of X.

Let us now consider the case where neither K nor D can be changed after the plant has been completed. The short-run total cost is defined by

$$SC(X, W, Q, K, D) = \inf \{WL + QE; Y \in T\}, \tag{5}$$

where once again the infimum is actually a minimum.

It easily shown that SC is concave in (W, Q) and convex in (X, K, D). The following relations are also easy to prove:

$$\partial SC/\partial W = L(X, W, Q, K, D), \partial SC/\partial Q = E(X, W, Q, K, D), \tag{6}$$

where $L(\cdot)$ and $E(\cdot)$ are the short-run demand functions for labor and energy, respectively. Moreover, the short-run supply of production is given by

$$\partial SC/\partial X = P, \tag{7}$$

where P is the output price. The relation between the short-run and long-run cost functions is given by

$$\inf_{K, D} (SC(X, W, Q, K, D) + RK) = TC(X, W, Q, R). \tag{8}$$

Let us now introduce uncertainty by assuming that Q is a random variable. As Q changes, so will the unit and marginal costs and therefore also the price of the output (unless demand is completely elastic). But as the price of output changes, so will the output itself. Thus X should also be treated as a random variable. Of course, X and Q are not independently distributed, so let their joint probability measure be $\mu(X, Q)$. The expectations of the two variables are denoted by $\bar{X}$ and $\bar{Q}$, respectively.

The problem of designing the plant can now be formulated as finding

$$\inf \int SC(X, W, Q, K, D)\,d\mu(X, Q) + RK, \tag{9}$$

where it is implicitly assumed that the plant manager is risk neutral. The solution to this problem yields the following planning functions (for simplicity we assume they are single valued):

$$K = K^u(W, R) \tag{10}$$

$$D = D^u(W, R). \tag{11}$$

These planning functions give the optimal capital stock and its optimal design as functions of the wage rate and the cost of capital when energy cost and output volume are random variables.

A very common rule of thumb when dealing with uncertainty is to replace uncertain variables with their expected values. The problem of risk then becomes deterministic. Such a replacement in this particular problem would imply that the design of the plant could be determined by

$$\inf SC(\bar{X}, W, \bar{Q}, K, D) + RK. \tag{12}$$

The solution to this deterministic problem is given by

$$K = K^c(\bar{X}, W, \bar{Q}, R) = \partial TC(\bar{X}, W, \bar{Q}, R)/\partial R \tag{13}$$

$$D = D^c(\bar{X}, W, \bar{Q}, R). \tag{14}$$

It is clear that (13) and (14) will provide a technically efficient plant design in the sense that the corresponding vector $(\bar{X}, L, E, K^c, D^c)$ will belong to $\bar{T}$. This is also the procedure we follow when using the simulation model.

However, it is by no means certain that the design corresponding to (10) and (11) will be technically efficient. All technically efficient points, i.e. all elements in $\bar{T}$, can be generated by parametric variation of $(\bar{X}, W, \bar{Q}, R)$ in (13) and (14). However, unless SC is linear in X and Q, (10) and (11) cannot be generated from (13) and (14), i.e. the solution to the uncertainty problem will not in general be technically efficient. The intuitive reason is obvious. By choosing an inefficient design, it is usually possible to obtain increased flexibility and elasticity with respect to changes in energy prices.

By definition then:

$$SC(\bar{X}, W, \bar{Q}, K^c, D^c) + RK^c < SC(\bar{X}, W, \bar{Q}, K^u, D^u) + RK^u \tag{15}$$

and

$$\int SC(X, W, Q, K^c, D^c)\,d\mu(X, Q) + RK^c > \int SC(X, W, Q, K^u, D^c)\,d\mu(X, Q) + RK^u. \tag{16}$$

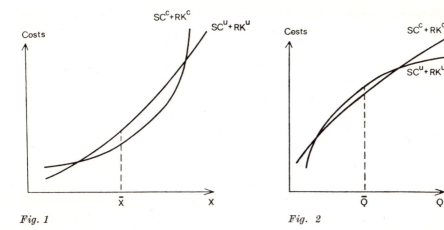

Fig. 1   Fig. 2

As SC is concave in Q and convex in X, it follows that unless the distribution of (X, Q) is quite odd, the relation between the solution to the certainty and risk problems may be described as shown in Figs. 1 and 2. These two figures indicate that for small deviations from the expected values of Q and X, the choices of K and D in the certainty case will yield smaller total expected costs than the optimal strategy for the uncertainty case, while large deviations will improve the optimal solution for the uncertainty case more *ex post* than the corresponding solutions for the certainty case. As the demand for energy will fall along with increases in its price, it follows that the solution to the uncertainty case will yield a more elastic demand for energy and a more elastic supply of output than the corresponding solutions for the certainty case.

Thus, the initial choice of capital stock and its design will be very important in influencing the social cost of changes in energy prices. If the capital stock of society is designed as cost minimizing for a particular set of prices, the cost of deviations in prices from this set may be much higher than if the stock were designed to accomodate price uncertainty. The primary purpose of this article is to try to quantify these aspects, using a simulation model of a general equilibrium type.

III. The Simulation Model

In order to carry out some numerical experiments along the lines suggested in the preceding section, we used a computable model[1] of the Swedish economy. The model is specified in accordance with basic notions of general equilibrium theory. It is designed for simulations of economic equilibria under various assumptions about factors such as world market conditions, domestic factor endowment and energy policy.

---

[1] The model is described in detail in Bergman & Por (1981).

In the basic version of the model presented below, both capital and labor are fully flexible factors of production and prices are flexible enough to clear all product and factor markets immediately. As will become apparent later on, some of the simulation experiments imply that constraints on capital utilization and labor mobility are imposed.

Consider an economy with $n+2$ production sectors, producing $n+2$ goods. There is no joint production and each good is produced in one sector only. The n first sectors produce tradable goods, which are also used for household consumption purposes. Sector $n+1$ is the housing sector and sector $n+2$ the public sector. There is also an additional "book-keeping" sector which produces an aggregated capital good, with sector index $n+3$.

There are two types of energy, fuels and electricity, produced by the first two sectors. In the long run fuels and electricity are similar to capital and labor in that they are variable factors of production, while other produced inputs are used in fixed proportion to output. The technology exhibits constant returns to scale in all sectors.

In general, imported and domestically produced units of goods with a given classification, say i, are imperfect substitutes from the users' point of view. This approach, introduced by Armington (1969), implies that domestic users of good i actually use a composite of imported and domestically produced units of goods with the classification[1] i. The composition of the composite good is determined on the basis of a preference function and cost minimization considerations. In this model, the preference function is assumed to be homothetic and to apply to all domestic users.

Assuming perfect competition, the prices of domestically produced goods, P_j, are equal to the unit production costs of these goods, $\varkappa_j$. Thus it holds that

$$P_j = \varkappa_j(P_1^D, ..., P_n^D, W_j, R_j); \quad j = 1, 2, ..., n+2 \tag{17}$$

where P_i^D is the price of composite good i and W_j the wage rate in sector j.[2] R_j is the user cost of capital in sector j, defined by[3]

$$R_j = P_{n+3}(\delta_j + R); \quad j = 1, 2, ..., n+2 \tag{18}$$

where P_{n+3} is the price of the aggregated capital good, δ_j is the rate of depreciation in sector j and R the real interest rate.

The unit cost function $\varkappa_j(\cdot)$ can be written

$$\varkappa_j(\cdot) = \varkappa_j^*(P_1^D, P_2^D, W_j, R_j) + \sum_{i=3}^{n} P_i^D a_{ij} + Q_j b_j; \quad j = 1, 2, ..., n+2 \tag{19}$$

---

[1] According to some commodity classification system such as the SITC.
[2] The heterogeneity of labor is roughly accounted for by an exogenous sectoral wage structure, i.e. $W_j = \omega_j W$ where ω_j is a constant.
[3] The price of capital goods is defined by $P_{n+3} = \sum_{i=1}^{n} P_i^D a_{i,n+3}$.

where the last two terms reflect the cost of nonsubstitutable inputs, nonenergy intermediate goods and complementary imports, respectively,[1] while $\varkappa_j^*(\cdot)$, the "net unit cost function", reflects the minimum cost of energy, labor and capital.

The net unit cost function, $\varkappa^*(\cdot)$, is derived from a nested Cobb–Couglas–CES production function, defined for the production sectors $j = 1, 2, ..., n+2$ by

$$X_j = A_j \{a_j F_j^{\varrho_j} + b_j H_j^{\varrho_j}\}^{1/e_j} e^{\lambda_j}$$

$$F_j = K_j^{\alpha_j} L_j^{1-\alpha_j}$$

$$H_j = \{c_j X_{1j}^{\gamma_j} + d_j X_{2j}^{\gamma_j}\}^{1/\gamma_j}$$

and the constraint

$$\lambda_j = \psi_j(\varrho_j, \gamma_j), \tag{20}$$

where X_j is gross output, K_j capital input, L_j labor input and X_{ij} ($i = 1, 2$) energy input. This production structure has three design parameters, i.e. the substitution parameters ϱ_j and γ_j[2] and the efficiency parameter λ_j. Thus, *ex ante* there is a trade-off between flexibility and efficiency; higher flexibility (i.e. lower absolute values of ϱ_j and γ_j) can be achieved at the expense of lower λ_j. The trade-off is given by eq. (20). This means that the design parameter D in the preceding section is represented by parameters ϱ_j, γ_j and λ_j in the simulation model. The design is assumed to be chosen according to the discussion in Section II. *Ex post* only labor and energy inputs are substitutable, and the values of ϱ_j, γ_j, λ_j and K_j are all fixed.

The equilibrium prices of the composite goods are given by the unit cost functions of the composites, that is by

$$P_i^D = \phi_i(P_i, P_i^M); \quad i = 1, 2, ..., n \tag{21}$$

where P_i^M is the exogenously given world market price, in the domestic currency unit, of import good i.

There are two types of domestic demand for composite goods, intermediate demand and final demand by the household sector. According to Shephard's lemma and the technology assumptions, the intermediate demand functions are given by

$$X_{ij} = \begin{cases} \dfrac{\partial \varkappa_j^*}{\partial P_i^D} \cdot X_j & \text{when} \quad i = 1, 2 \\ a_{ij} X_j & \text{when} \quad i = 3, 4, ..., n \end{cases} ; \quad j = 1, 2, ... n+3 \tag{22}$$

---

[1] Thus, a_{ij} is the input of composite good i per unit of output in sector j, and $\bar{b}_j$ is the corresponding parameter for complementary imports. The world market price of complementary imports, expressed in the domestic currency unit, is Q_j.

[2] Note that ϱ_j and γ_j are the substitution parameters. The corresponding substitution elasticities are given by $1/(1-\varrho_i)$ and $1/(1-\gamma_j)$.

120 L. Bergman and K.-G. Mäler

Household demand is given by

$$C_i = C_i(P_1^D, ..., P_{n+1}^D, E); \quad i=1, 2, ..., n+1 \tag{23}$$

where E denotes total household consumption expenditures. In the model, eqs. (23) are represented by a linear expenditure system estimated on ten consumer commodity groups, each defined as a convex combination of composite goods.

Foreign demand for domestically produced goods is given by the export functions

$$Z_i = Z_i(P_i, P_i^W, t); \quad i=1, 2, ..., n \tag{24}$$

where P_i^W is the exogenously given world market price of goods with classification i, and t is a time index.

On the basis of Shephard's lemma and eqs. (17)–(24), the equilibrium conditions for the product markets can be written

$$X_i = \frac{\partial \phi_i}{\partial P_i} \left\{ \sum_{j=1}^{n+3} X_{ij} + C_i \right\} + Z_i; \quad i=1, 2, ..., n \tag{25}$$

$$X_i = C_i; i = n+1, n+2 \tag{26}$$

$$X_{n+3} = I + \sum_{j=1}^{n+2} \delta_j \frac{\partial \varkappa_j^*}{\partial R_j} X_j, \tag{27}$$

where C_{n+2} is exogenously given public consumption and I is exogenously given net investments.

The demand for competitive imports is given by

$$M_i = \frac{\partial \phi_i}{\partial P_i^M} \left\{ \sum_{j=1}^{n+3} X_{ij} + C_i \right\}: \quad i=1, 2, ..., n \tag{28}$$

and current account equilibrium implies

$$\sum_{i=1}^{n} P_i Z_i = \sum_{i=1}^{n} P_i^M M_i + \sum_{i=1}^{2} Q_j \bar{b}_j X_j. {}^{1} \tag{29}$$

Similarly, assuming that capital and labor are inelastically supplied, the equilibrium conditions for the factor markets become

$$K = \sum_{j=1}^{n+2} \frac{\partial \varkappa_j^*}{\partial R_j} X_j \tag{30}$$

$$L = \sum_{j=1}^{n+2} \frac{\partial \varkappa_j^*}{\partial W_j} X_j. \tag{31}$$

[1] Observe that complementary imports are used in the energy sectors only. These inputs are thus imported energy resources, i.e. crude oil in the refinery ector and coal or uranium in the electricity sector.

As P_{n+3} is determined by P_i^D, $i=1, 2, ..., n$, and all sectoral wage rates W_j are proportional to a general wage rate, W, there are $6n+9$ equations in the $6n+9$ unknowns $X_1, ..., X_{n+3}$, $C_1, ..., C_{n+1}$, $Z_1, ..., Z_n$, $M_1, ..., M_n$, $P_1, ..., P_{n+2}$, $P_1^D, ..., P_n^D$, E, W and R.

In the model outlined above, changes in the world market prices of energy resources, i.e. Q_1 and Q_2, will transmit themselves into product price increases through the cost functions (19). Assume that, initially, the world market prices of energy resources are given by the vector Q^0, and the corresponding vector of product prices is P_0^D. Now, let energy resource prices change to Q^1, thereby leading to he equilibrium product price vector P_1^D. If the initial level of utility is taken to be u^0, the value of the expenditure function in the initial price system is $m(P_0^D, u^0)$, while the corresponding value after the energy price increase is $m(P_1^D, u^1)$.

The cost to society will be measured by the quivalent variation, EV, implicitly defined by

$$v(P_0^D, E_0 + EV) = v(P_1^D, E_1) = u^1, \qquad (32)$$

where E_0 and E_1 are the household expenditure levels in the two situations and $v(\cdot)$ is the indirect utility function. The difference between E_0 and E_1 reflects changes in producers' surplus and the wage rate. As $E_0 + EV = m(P_0^D, u^1)$, the EV is defined by

$$EV = m(P_0^D, u^1) - E_0 \approx P_0^D C_1 - E_0, \qquad (33)$$

where C_1 is the household consumption vector when prices are P_0^D. The approximation indicated in (33) is used in the numerical calculations.

It may seem that the general equilibrium framework has little to do with our previous discussion of efficiency versus flexibility. The time has now come to attempt to integrate these two discussions.

The production structures in the simulation model contain three design parameters, the substitution parameters ϱ_j and γ_j, and the efficiency parameter λ_j. We have assumed that the plant manager can choose these parameters *ex ante*, subject only to the restriction (20), i.e.

$$\lambda_j = \psi_j(\varrho_j, \gamma_j), \qquad (20)$$

where ψ_j is an increasing function of its two variables. However, we had no possibility of estimating this function and were therefore forced to neglect the cost of achieving increased *ex post* flexibility. Instead, we concentrate on the benefits from increases in substitution elasticities. Our situation can most easily be explained in Figure 3 where, for simplicity, only fuels are considered. This diagram corresponds to Figure 2, but with the important difference that all cost curves in Figure 3 are tangent to each other at the expected price $\overline{Q}$.

---

[1] In this context we disregard distributional considerations.

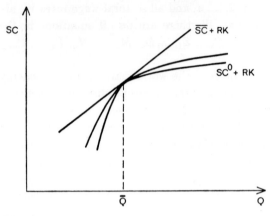

Fig. 3

The different cost curves correspond to different values of the design parameters. The curve $\bar{S}\bar{C}$ corresponds to the case where the production structure
is completely rigid *ex post*, i.e. fixed input coefficients and substitution elasticities equal zero. SC^0, on the other hand, corresponds to the case where the
ex post substitutability corresponds to a Cobb Douglas production function,
i.e. the numerical values of the substitution elasticities are close to unity. This
is the most "flexible" case we consider. The efficiency parameter would in
general vary among these different sets of substitution parameters according
to the function $\psi_j(\cdot)$. As we could not estimate this function, we have choosen
λ_j so that each cost curve is tangent to the *ex ante* cost curve at the expected
price $\bar{Q}$. By doing this we essentially disregard the cost of increased flexibility,
which would show up as vertical differences between the different curves at
the expected price. The vertical difference between the curves in Fig. 3 at a
particular price Q' can be interpreted as the gross benefit in the respective
sector of increases in *ex post* flexibility if price Q' is realized.

Thus, the simulation model may be calibrated in such a way that at a
predetermined price vetctor, interpreted as the expected price vector, the model
will yield the same results independent of the sssumptions about the *ex post*
substitution elasticities, i.e. the design parameters. By then assuming that
other prices are realized, it may be possible to compute the gain to society, as
measured by EV, from this increased flexibility.

IV. Some Numerical Experiments

The main purpose of the experiments is to obtain a rough estimate of the social
benefits of a flexible *ex post* technological design when energy resource prices
attain unexpected values. In theory, we would have adhered to the discussion
in Section II and posited a probability distribution over future energy prices.

Table 1. *The production sectors*

Number	Sector
1	Fossil fuels production[a]
2	Electricity production
3	Mainly import-competing industries
4	Mainly exporting energy-intensive industries
5	Other mainly exporting industries
6	Sheltered industries and service production
7	Public sector
8	Capital goods sector (book-keeping sector)

[a] Almost entirely oil refining.

Then, for each probability distribution, we would have measured the corresponding expected benefits. Instead, we will deal only with the benefits corresponding to a particular outcome. Obviously, there is an element of arbitrariness in such estimates. For instance, there is an almost infinite number of possible energy resource price constellations, and the value of our cost measure may differ considerably between them. Moreover, we consider only a few *ex post* technology designs. We do not know the probability that these designs would be chosen under reasonable assumptions about the trade-off between efficiency and flexibility and about the subjective probability distributions of energy resource prices.

However, the simulation model can be used to compare the economic significance of an inflexible *ex post* technology and other rigidities which constrain the economy's adjustment to changes in energy resource prices. Thus, even in view of the above qualifications, the quantitative importance of the efficiency-flexibility trade-off can be indicated on the basis of the model simulations.

For this study the model was implemented on an eight sector data base. The main data source is an input–output table for 1975, aggregated to the eight-sector level. Extraneous information about the parameters in import, export and household demand functions has also been used.[1] The sector classification is presented in Table 1.

The actual energy resource prices in 1975 are taken to be equal to expected energy resource prices. For given values of the efficiency parameters $\lambda_1, ..., \lambda_7$, the sectoral *ex ante* production functions are all assumed to be approximately Cobb–Douglas in capital, labor, fuels and electricity. On the basis of these *ex ante* production functions, the expected values of energy resource prices and the actual 1975 values of other exogenous variables, a full general equilibrium is computed.

---

[1] A full description of the data base is given in Bergman & Mäler (1981).

Table 2. *The* ex post *designs*

h	$(1 - \varrho(h))^{-1}$	$(1 - \gamma(h))^{-1}$
1	~ 0	~ 0
2	0.6	0.2
3	0.2	0.6
4	~ 1.0	~ 1.0

In accordance with the discussion at the end of the preceding section, we assume, in effect, that all the *ex post* production functions considered exhibit the same efficiency when energy resource prices attain their expected values. Thus, regardless of the design of the *ex post* technology, the "short-run" equilibria consistent with each of the *ex post* technologies coincide at these energy resource prices. The utility level (approximated by the consumption vector) attained at this equilibrium is taken as the standard of comparison.

Four different *ex post* technology designs are considered. None of them contain any differentiation between sectors in terms of the substitution parameters ϱ_j and γ_j. Accordingly, the sector indices on these parameters can be dropped, and each design $D(h)$ is characterized fully by $\varrho(h)$, $\gamma(h)$. The characteristics of the four *ex post* designs in terms of the implied elasticities of substitution are summarized in Table 2.

Thus $D(1)$ implies an *ex post* technology which is rigid in terms of energy use per unit of output in each sector. $D(4)$, on the other hand, implies unitary substitutability between energy and other factors of production, as well as between fuels and electricity. Design $D(2)$ is relatively flexible in terms of aggregate energy use, although less so than $D(4)$, and relatively rigid in terms of the composition of aggregated energy, but less so than $D(1)$. As compared to $D(2)$, design $D(3)$ has the opposite properties.

Under the assumption that the four designs are equally efficient when energy resource prices attain their expected values, $D(1)$ is clearly an inferior design, while $D(4)$ is superior to the others. However, these two designs are incorporated only in order to obtain estimates of reasonable upper and lower bounds on the cost associated with a deviation from expected energy resource prices.

Designs $D(2)$ and $D(3)$ represent cases where uncertainty about future prices leads to a choice of design which allows for some flexibility *ex post*. However, in order to represent an optimum choice of *ex post* technology, these two designs reflect different expectations about future energy prices. Thus, $D(2)$ reflects expectations about rather uniform future energy price changes, while $D(3)$ reflects expectations about changes in the price relations between fuels and electricity. The ranking of $D(2)$ and $D(3)$ obviously depends on the actual development of energy prices.

Generally the use of capital is fixed *ex post*. In the simulation model, this

Table 3. *Estimated cost of an 80% increase in crude oil prices under full short-run adjustment*

h	$EV(h)^a$	$(EV(h)/E^0)100$	$(EV(h)/EV(1))100$	$EV(h) - EV(1)^b$
1	$-7\ 590$	-5.0	100	0
2	$-7\ 020$	-4.6	92.5	570
3	$-6\ 920$	-4.5	91.2	670
4	$-5\ 260$	-4.0	80.7	2 330
$1/\bar{L}$	$-11\ 380$	-7.4	150.0	$-3\ 790$

[a] Expressed in 10^6 Sw. Kr. at the 1975 price level.
[b] This is a measure of the benefit of higher flexibility (in $h = 1, ..., 4$) and labor market rigidities ($h = 1/\bar{L}$).

means that the equilibrium condition for the capital market, i.e. eq. (30) in Section III, has to be deleted; the same applies to the variable R, the real interest rate. Moreover, the "user cost of capital" variables R_j have to be replaced by a set of quasi-rents, Π_j, defined as the difference between gross receipts and variable costs. Although the model cannot readily be changed in this way, it turned out to be relatively easy to find a set of Π_j which kept the sectoral use of capital at a constant level, in spite of changes in energy resource prices.

The analysis was focused on a case where the world market price of crude oil, i.e. the variable Q_1 in the simulation model, increases to a level 80% higher than the expected level. All prices are assumed to attain their new equilibrium values instantaneously and aggregate demand in the "rest of the world" is assumed to be kept at the full employment level. However, even under these conditions the energy price increase entails a social cost. Estimates of this cost are presented in Table 3.

Clearly, the real income loss due to an oil price increase of the magnitude assumed is significant, even when the technology is as flexible as in $D(4)$ and a new equilibrium is instantaneously attained. However, the degree of flexibility certainly matters; as compared to the rigid design $D(1)$, flexible *ex post* technologies such as $D(2)$ or $D(3)$ reduce the real income loss by about 8%. Accordingly, society benefits from flexible technology if the price of oil attains an unexpectedly high value.

The results in the last row of Table 3 were obtained for a case where the assumption about full adjustment was relaxed. It was assumed that employment could not instantaneously increase in any sector, while immediate reductions in sectoral employment would still be possible. The technology was assumed to be $D(1)$. As the equilibrium solution implied increased employment in two sectors,[1] the additional constraint would lead to 2.4% overall unemployment.

---

[1] Sectors 3 and 5.

Table 4 A. *Estimates of $EV(h)^a$ for different technology designs at selected oil price levels*

	Q_1		
h	1.00^b	1.80	2.60
1	0	− 7.590	− 14 790
5	− 2 360	− 8 350	− 13 040

^{a} Measured in 10° Sw. Kr. at the 1975 price level.
^{b} Equal to the expected value for Q_1.

The results indicate that even if short-run rigidities in the technology account for a significant part of the real income loss due to an unexpected oil price increase, rigidities in the functioning of factor markets can be of much greater importance. Other types of indirect effects of an oil price increase, leading to a drop in aggregate demand, can be even more significant. Thus, with reasonable assumptions about which *ex post* flexibility can be attained, flexibility-increasing energy policies may not be the most important way of reducing the economy's vulnerability to unexpected oil price increases.

However, the results of our calculations indicate that significant value can be assigned to technologies with flexible *ex post* input proportions. As policies aimed at increasing the *ex post* flexibility of technology do not necessarily constrain the possibilities of pursuing policies aimed at improving the functioning of factor markets or at stabilizing aggregate demand, it seems as if a case can be made for some emphasis on flexibility-increasing measures in energy policy. However, the optimum degree of flexibility cannot be determined without an explicit specification of the function $\psi_j(\cdot)$ defined by eq. (20) in Section III.

In order to illustrate the value of *ex post* flexibility in a slightly different way, we made the following assumptions

(i) $\psi_j(\cdot) = \psi(\cdot)$ $j = 1, 2, ..., 7$

(ii) $-0.01 = \psi(0, 0)$

(iii) $0 = \psi(-\infty, -\infty)$.

Thus, it was assumed that a design implying an *ex post* technology which is Cobb–Douglas in capital, labor, fuels and electricity would be 1 % less efficient in these factors than design $D(1)$. If energy resource prices attain their expected values when a design with these characteristics is chosen, the model calculations indicate that the economy's equilibrium would differ from the equilibrium with design $D(1)$ in two ways. First, the real income level would be 1.5 % lower. Second, the use of oil per unit of output in the production sectors would be

Table 4B. *Estimates of $(EV(h)/E^0)100$ for different technology designs at selected oil price levels*

h	Q_1		
	1.00	1.80	2.60
1	0	-5.0	-9.7
5	$\leqslant 1.5$	-5.5	-8.5

higher. For instance, in sector 4, the most energy-intensive sector, the use of oil per unit of output is 1.9 % higher than the corresponding value under $D(1)$.

The interesting point, however, is what happens if oil prices attain an unexpectedly high level. Table 4 and Fig. 4 summarize the comparison of $D(1)$ and the flexible, somewhat less efificient design which, for simplicity, is called $D(5)$. Fig. 4 shows the results in terms of the theoretical framework developed in Section II. Note that the cost curves in Section II applied to an individual production unit, while the curves in Fig. 4 essentially reflect the "welfare cost" for society as a whole.

The simulation results show that the value of *ex post* flexibility depends on the range of possible deviations of prices from their expected values. This result was also indicated in Fig. 2. Thus, under the assumptions applied in this analysis, it would not pay to choose $D(5)$ rather than $D(1)$ if oil prices are highly unlikely to deviate from expected levels by more than 80 %. If, on

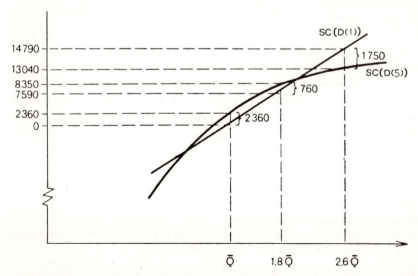

Fig. 4. Costs and benefits from increased *ex post* flexibility. SC(D(1)) corresponds to total cost with rigid *ex post* input/output coefficients and SC(D(5)) corresponds to *ex post* Cobb-Douglas technology.

the other hand, deviations of 160 % or more are possible or even likely, $D(5)$ can be the best choice. Again, however, the optimum choice depends on the exact specification of the probability distribution for Q_1.

It has already been pointed out that $D(5)$ implies higher oil consumption per unit of output than $D(1)$ when $Q_1 = 1.00$, At $Q_1 = 2.60$, however, the situation is radically different. While oil input coefficients are essentially unchanged under $D(1)$, they are about 50 % lower under $D(5)$. This result illustrates that an emphasis on energy conservation at the expense of short-term flexibility in energy use may not be the best way of reducing the economic risks associated with unexpected oil price increases.

V. Conclusions

Although our numerical simulations are somewhat arbitrary, a major conclusion is that investments in increased flexibility may yield substantial benefits if uncertainty about future fuel prices is sufficiently high. As we were not able to estimate the cost of increased flexibility, all of our quantitative results are of course arbitrary. However, the possibility of reducing the welfare costs of substantial increases in future fuel prices by more than a billion. Sw. Kr. is sufficiently attractive to motivate further research in this field. A next step should then be to find techniques for estimating the cost, in terms of efficiency reductions, of various kinds and degrees of *ex post* flexibility. This would make it possible to carry out a real cost-benefit study of efficiency versus flexibility.

Another noteworthy conclusion is the importance of mobility on the labor market. If the labor market is completely rigid, then the cost of an 80 % increase in fuel prices would be Sw. Kr. 4 billion more than if labor could be freely reallocated among sectors. In view of the highly aggregated sectors we have used in the simulation model, this may even be an underestimate. In any case, the result indicates a need for further research on the role of mobility on the labor market in terms of accommodating unexpected pricei ncreases. In all simulations reported in this study, capital stocks have been locked in each sector. But we have also carried out simulations in which capital can move freely between different sectors. In contrast to the case of labor, it does not seem to matter very much whether capital is mobile or not.

References

Armington, P. S.: A theory of demand for products distinguished by place of production. *IMF Staff Papers 16*, 159–178, 1969.

Bergman, L. & Mäler, K.-G.: Förväntningsbildning och effekten av höjda olje-

priser (Expectation formation and the impact of higher oil prices), forthcoming, 1981.

Bergman, L. & Por, A.: *Computable Models of General Equilibrium in a Small Open Economy*, forthcoming, 1981.

EMPLOYMENT EFFECTS OF AN INCREASED OIL PRICE IN AN ECONOMY WITH SHORT-RUN LABOR IMMOBILITY

Michael Hoel

University of Oslo, Oslo, Norway

Abstract

The effects of an increased oil price are analyzed in a two-sector economy with immobile labor combined with a "wage leadership" assumption. Starting with a full employment equilibrium, an increased oil price will generally lead to unemployment in one of the sectors. This unemployment may be eliminated by an appropriate change in fiscal policy. If unemployment occurs in the sector which produces nontradeables, taxes should be reduced whereas taxes should be increased if unemployment occurs in the sector which produces tradeables. If an increased oil price leads to an international excess supply of tradeable goods, the direction of the tax change necessary to eliminate unemployment is unambiguous even when the sector in which there is unemployment is known.

I. Introduction

There is a large and growing literature which treats the trade balance and employment effects of rising oil prices; see e.g. Cooper (1976), Gordon (1975), Gramlich (1979), Hoel (1979), Phelps (1978), Pindyck (1979), Rødseth (1978) and Solow (1978). All of this literature focuses on a one-sector economy. The reason why higher oil prices may lead to increased unemployment is that there is some kind of wage and/or price stickiness in the economy.

The effects of increased oil prices may of course also be analyzed within the framework of a two-sector economy which produces tradeables and nontradeables; see e.g. Steigum (1980). However, in practically all such two-sector models used for short-run macroeconomic analysis, perfect labor mobility is assumed between the two sectors. This is clearly an unrealistic assumption in the short run. This paper therefore considers the opposite extreme, i.e. that there is *no* labor mobility between the two sectors. In order to concentrate on this aspect of the economy, we disregard domestic price and wage rigidity in our analysis. The wage rate, however, will be determined in a somewhat unusual manner. When there is no labor mobility we have two completely separate labor markets. A Walrasian equilibrium would then normally have different equilibrium wages in the two labor markets. In most modern economies, however, we expect there to be a close connection between wage rates in different sectors, in spite of short-run labor immobility. We include this wage rate interdependence in our model by a simple "wage leadership" assumption: the wage rate is transmitted from the "leading sector" to the rest of the economy. A consequence of this assumption is that we always have the same wage rate in the two labor markets. The justification for this assumption is that local wage settlements are strongly influenced by wage rates elsewhere in the economy, and that "fairness" considerations are important in wage rate determination. Further justification for this type of transmission mechanism has been provided by e.g. Hicks (1975), Kaldor (1959) and Scitovsky (1978 a, b).

In our model the "leading sector" at any time is simply the sector which has the highest market-clearing wage rate. The justification for this assumption is that in the absence of wage rigidity, it is difficult to imagine the possibility of excess demand for labor in a market. "Fairness" considerations will seldom prevent local wage settlements from resulting in a wage rate *higher* than the wage rate elsewhere in the economy. Our wage determination must therefore imply that wage rates are always equal in both sectors, and equal to the market-clearing wage rate of the sector in which this wage rate is highest.

A simple model which incorporates the features above is presented in Section II. The effects of the increased oil prices are discussed in Section III. Policy implications are summarized in Section IV.

II. The Model

The model consists of the following relations, which will be explained below:

$$X_1 = D_1(\underset{?}{p}, \underset{+}{Y}) + G_1 + E, \tag{1}$$

$$X_2 = D_2(\underset{-}{p}, \underset{+}{Y}) + G_2, \tag{2}$$

$$Y = (1-t)[X_1 + pX_2 - qR], \quad 0 < t < 1, \tag{3}$$

$$R = R(\underset{-}{q}, \underset{+}{w}, \underset{+}{X_1}, \underset{+}{X_2}), \tag{4}$$

$$Z = t[X_1 + pX_2 - qR] - [G_1 + pG_2], \tag{5}$$

$$X_1 = S_1(\underset{-}{w}, \underset{-}{q}), \tag{6a}$$

$$X_1 = A + k[D_1(p, Y) + G_1], \quad 0 < k < 1, \tag{6b}$$

$$X_2 = S_2(\underset{+}{p}, \underset{-}{w}, \underset{-}{q}), \tag{7}$$

$$L_1(\underset{-}{w}, \underset{-}{q}) \leqslant L_1, \tag{8a}$$

$$L_1^c(\underset{-}{w}, \underset{+}{q}, \underset{+}{X_1}) \leqslant L_1, \tag{8b}$$

$$L_2(\underset{+}{p}, \underset{-}{w}, \underset{-}{q}) \leqslant L_2. \tag{9}$$

Relations (1) and (2) show that the production of tradeables and nontradeables (X_1 and X_2) is used for private expenditure (D_1 and D_2), public expenditure (G_1 and G_2) or net export (E for tradeables, zero for nontradeables). The domestic currency price of tradeables is set equal to 1 by choice of units, so that p, q and w stand for the price of non-tradeables, the oil price and the wage rate, all in terms of the tradeable. D_1 and D_2 are standard demand functions satisfying $D_1 + pD_2 = Y$, with signs of partial derivatives indicated under the variables in (1) and (2). Private disposable income Y is given by (3), where t is the tax rate and R is the amount of oil used. For simplicity, we disregard direct use of oil, so that oil demand is given as an intermediate (normal) input by (4). When $D_1 + pD_2 = Y$, the trade surplus ($= E - qR$) must simply equal the government budget surplus ($= Z$); cf. (5).

When there is no demand constraint, the supply functions are given by (6a) and (7), where $S_2(p, w, q)$ is homogeneous of degree zero in p, w and q, where both inputs are normal. Since we assume that the price of nontradeables is completely flexible, the case of a demand constraint for the producers of non-tradeables is irrelevant. However, we consider the possibility of an increase in the oil price leading to an international excess supply disequilibrium situation for the tradeable good. In such a situation the producers of tradeables will face a demand constraint. The assumption underlying (6b) is that this demand consists of exogenous exports (A) and an exogenous share (k) of domestic demand. This treatment of an international excess supply situation is identical to the approach of Dixit & Norman (1980).

With no demand constraints, labor demand is given by (8a) and (9), while (8b) must be substituted for (8a) in the case of an international excess supply of tradeables. Labor and oil are assumed to be complementary in the unconditional demand functions; see Hoel (1980a) for a justification. A consequence of this assumption is that the conditional demand function $L_c^1(\cdot)$ must satisfy

$$\partial L_1^C / \partial q + (\partial L_1^C / \partial X_1)(\partial S_1 / \partial q) < 0.$$

The labor supply in each sector is exogenous (L_1 and L_2). Our assumption regarding wage rate determination implies that at least one of the constraints L_1 and L_2 are binding.

The model consists of 8 equations which determine X_i, w, p, Y, E, R, Z with G_i, T, q, A, k and L_i given exogenously. Questions regarding the existence or uniqueness of a solution will not be discussed; we simply assume that the functions have properties which imply a unique solution.

The main concern of this paper is to analyze the effects of an increased oil price on the two labor markets. For this purpose it is convenient to derive reduced form labor demand functions where the explanatory variables are w and all the exogenous variables. Let us start with the case in which the world market for tradeables remains in equilibrium after an oil price increase, so that

(6a) and (8a) are the relevant equations. In this case the relative price of non-tradeables (p) is determined from (2), considering (3), (4), (6a) and (7):

$$D_2(p, (1-t)[S_1(w, q)+pS_2(p, w, q)-qR(q, w, S_1(w, q), S_2(p, w, q)]+G_2$$
$$=S_2(p, w, q).$$

We assume that the excess demand for nontradeables is a declining function of the price of nontradeables, i.e. $d(D_2(\cdot)-S_2(\cdot))/dp<0$ (a *sufficient* condition for this is that $D_{2p}+(1-t)X_2D_{2Y}<0$). In this case the expression above gives us

$$p = p(\underset{?}{w}, \underset{?}{q}, \underset{+}{G_2}, \underset{-}{t}). \tag{10}$$

The reason why $p_w<0$ and $p_q<0$ are possible is that both oil price and wage increases will have a negative income effect, thereby tending to reduce p.

Inserting (10) into (9) gives us the reduced form demand for labor in the sector which produces nontradeables:

$$L_2^R = L_2^R(\underset{?}{w}, \underset{?}{q}, \underset{+}{G_2}, \underset{-}{t}). \tag{11}$$

In the subsequent analysis we assume that the absolute value of $\partial L_2(p, w, q)/\partial w$ is sufficiently high to make $\partial L_2^R/\partial w<0$. The reduced form demand for labor in the tradeables sector is simply given by (8a), so that we obtain the following two inequalities in the case of no demand constraint:

$$L_1(\underset{-}{w}, \underset{-}{q}) \leqslant L_1, \tag{12}$$

$$L_2^R(\underset{-}{w}, \underset{?}{q}, \underset{+}{G_2}, \underset{-}{t}) \leqslant L_2, \tag{13}$$

where w is determined so that we get an equation sign in at least one of expressions (12) and (13).

Turning now to the case in which (6b) and (8b) are valid, p and X_1 are determined simultaneously by

$$D_2(p, (1-t)[X_1+pS_2(p, w, q)-qR(q, w, X_1, S_2(p, w, q)])+G_2 = S_2(p, w, q)$$

and

$$X_1 = A + k\{D_1(p, (1-t)[X_1+pS_2(p, w, q)-qR(q, w, S_2(p, w, q)])+G_1\}.$$

These equations give

$$p = \tilde{p}(\underset{?}{w}, \underset{?}{q}, \underset{+}{A}, \underset{+}{k}, \underset{+}{G_1}, \underset{+}{G_2}, \underset{-}{t}) \tag{14}$$

and

$$x_1 = \tilde{x}_1(\underset{?}{w}, \underset{?}{q}, \underset{+}{A}, \underset{+}{k}, \underset{+}{G_1}, \underset{+}{G_2}, \underset{-}{t}). \tag{15}$$

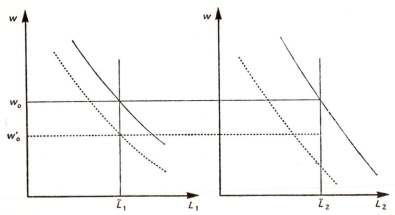

Fig. 1

The signs of the derivatives are valid for $d(D_2(\cdot)-S_2(\cdot))/dp<0$ (when X_1 as a function of p is inserted into $D_2(\cdot)$) and $D_{1p}+(1-t)(S_2+(p-q\partial R/\partial X_1)\partial S_2/\partial p>0$. I have further discussed conditions similar to these elsewhere (Hoel, 1980b); in what follows we simply accept the signs indicated in (14) and (15). As before, the possibility of $\tilde{p}_w<0$ and $\tilde{p}_q<0$ are due to a negative income effect on the demand for nontradeables when w or q increases.

Inserting (14) and (15) into (8b) and (9) gives us the reduced form demand functions for labor ($\tilde{L}_1^R(\cdot)$ and $\tilde{L}_2^R(\cdot)$). As in the previous case, we assume that the reduced form labor demand functions are declining in w, so that in the case of international excess supply of tradeables we obtain

$$\tilde{L}_1^R(\underset{-}{w}, \underset{?}{q}, \underset{+}{A}, \underset{+}{K}, \underset{+}{G_1}, \underset{+}{G_2}, \underset{-}{t}) \leqslant L_1,$$ (16)

$$\tilde{L}_2^R(\underset{-}{w}, \underset{?}{q}, \underset{+}{A}, \underset{+}{k}, \underset{+}{G_1}, \underset{+}{G_2}, \underset{-}{t}) \leqslant L_2,$$ (17)

where w is determined so that we get an equation sign in at least one of expressions (16) and (17).

III. The Effects of Increased Oil Prices

In this section we analyze how employment is affected by an increased oil price, and how the government must change the tax rate in order to restore full employment. We assume that initially, the economy is in a full employment equilibrium. The justification for this, of course, is that the immobility of labor is a typical short-run phenomenon; in a long-run equilibrium L_1 and L_2 will have adjusted so that (12) and (13)—(or (16) and (17))—both hold as equations.

Consider first the case in which there is no demand constraint for tradeables. The initial full employment equilibrium is illustrated by the heavily drawn curves in Fig. 1 (which are derived from (12) and (13) as two equations),

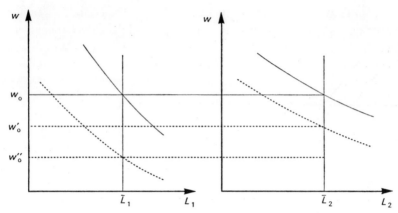

Fig. 2

giving w_0 as the equilibrium wage rate. An increased oil price will shift $L_1(w, q)$ to the left, as illustrated by the broken line in the left half of Fig. 1. As for $L_2^R(\cdot)$, there are three possibilities. The first is that $L_2^R(\cdot)$ is shifted to the left by a vertical distance (at $\bar{L}_2$) larger than the vertical distance of the shift in $L_1(\cdot)$ (at $\bar{L}_1$). This case is illustrated in Fig. 1. The new wage rate will be w_0', giving unemployment in sector 2. Since $L_1(w, q)$ is unaffected by fiscal policy, this unemployment can only be eliminated by raising $L_2^R(\cdot)$ up to the intersection of w_0' and $\bar{L}_2$. This can be achieved either by reducing the tax rate or by increasing government consumption of nontradeables. Since government consumption is to a large extent determined by considerations other than demand regulation, we focus on changes in the tax rate in the following analysis.

The second possibility is illustrated in Fig. 2. As in the case above, $L_2^R(\cdot)$ shifts to the left. In the present case, however, the shift in $L_2^R(\cdot)$ is smaller than the shift in $L_1(\cdot)$ (both measured vertically at $\bar{L}_2$ and $\bar{L}_1$, respectively). As before, the equilibrium wage shifts downwards from w_0 to w_0', but we now get unemployment in sector 1. In order to obtain full employment, $L_2^R(\cdot)$ must be shifted down to the intersection of w_0'' and $\bar{L}_2$. This requires an *increased* tax rate (or a reduction in G_2).

In both of these cases, the real wage rate (in terms of the tradeable good) declined as a consequence of the increased oil price. If we had assumed instead that this wage rate was rigid (at least downwards), unemployment in sector 1 could not be prevented. However, unemployment in sector 2 could be eliminated by raising $L_2^R(\cdot)$ back to its initial position. This would require a tax *reduction* in both of the cases above. It is therefore clear that the correct fiscal policy for reducing unemployment caused by an increased oil price depends heavily on wage determination in the economy.

The third possibility for $L_2^R(\cdot)$ is an upward shift. This case shares all of the properties of the case illustrated by Fig. 2, except that the w_0' will now lie above w_0.

We now turn to the case in which the increased oil price leads to a demand constraint for tradeables. In this case (16) and (17) must be used in our analysis, with equation signs in the initial full employment equilibrium. If there is no demand constraint prior to the oil price increase, we must let A initially have a value which makes $\tilde{L}_1^R(\cdot) = L_1(\cdot)$, implying $\tilde{L}_2^R(\cdot) = L_2^R(\cdot)$. The rise in the oil price now makes q increase and A decline, and the effects of these changes can be analyzed by figures similar to Figs. 1 and 2, with $L_1(\cdot)$ and $L_2^R(\cdot)$ replaced by $\tilde{L}_1^R(\cdot)$ and $\tilde{L}_2^R(\cdot)$. It is clear that the combined effects of increased q and reduced A on $\tilde{L}_i^R(\cdot)$ will be ambiguous if $\partial \tilde{L}_i^R(\cdot)/\partial q > 0$, although it seems reasonable to expect both $\tilde{L}_1^R(\cdot)$ and $\tilde{L}_2^R(\cdot)$ to shift to the left, as in Figs. 1 and 2 for the case without any demand constraint. As in the previous case, the sector in which we will get unemployment is not known *a priori*. In any case, in order to eliminate whatever unemployment there is, the tax rate must be adjusted so that $\tilde{L}_1^R(\cdot)$ and $\tilde{L}_2^R(\cdot)$ cut $\tilde{L}_1$ and $\tilde{L}_2$ at the same wage rate. Unlike the case without any demand constraint, however, knowledge of which sector has unemployment is not sufficient to determine the direction in which the tax rate should be changed. The reason for this ambiguity is, of course, that with a demand constraint the tax rate affects labor demand in *both* sectors.

IV. Concluding Remarks

Our analysis has shown that an increase in oil prices should not necessarily lead to a tax reduction (or other expansionary fiscal or monetary policies) even if the only concern is the level of unemployment. Our model gives two cases in which a tax reduction after an increase in the oil price will reduce unemployment, namely when the increased oil price leads to unemployment (i) in both sectors due to sticky wages and (ii) in the sector which produces nontradeables when there is no demand constraint for tradeables. If the increased oil price leads to unemployment in the sector which produces tradeables and there is no demand constraint for tradeables, then the tax rate must be *increased* to reduce unemployment. As for the other cases we have treated, the effect of a change in the tax rate on unemployment is ambiguous.

The reason for the ambiguity concerning how the tax rate should be changed to increase employment lies in our assumption about short-run labor immobility between sectors. An increased oil price will normally change the relationship between the market-clearing wage rates of the two sectors. If the actual wage rates do not adjust accordingly, we get unemployment. The only way this unemployment can be eliminated is to change the labor demand in one or both sectors in an appropriate way. Clearly, the income tax rate is a very crude policy tool for regulating the *composition* of labor demand. It is therefore not surprising that it is difficult to give any unambigous rule about how the tax rate should be changed.

One might well ask if changes in the tax rate *ought* to be used as a means of reducing unemployment in an economy where the main features of the present model play an important role. Since the *composition* of demand is so important in such an economy, it would probably be preferable to use more sector specific instruments than the tax rate to regulate the unemployment rate. The tax rate could then be used to regulate the trade balance, which has been ignored in this analysis.

Another question which is beyond the scope of the present paper is whether any efforts should be made to eliminate the unemployment which occurs after a rise in the oil price. Such unemployment is after all a consequence of short-run labour immobility, so that it will only be temporary if labor is mobile in the long run. If this temporary unemployment were eliminated, the incentives for a gradual reallocation of labor would also be removed, so that long-run resource allocation may be affected by the policies used to prevent temporary unemp-ployment.

References

Cooper, R. N.: The global impact of the oil price rise. In *World monetary disorder* (ed. P. M. Boardman and D. M. Tuerck), pp. 143–152. Praeger Publishers, New York, 1976.

Dixit, A. & Norman, V.: *Theory of international trade.* Cambridge University Press, 1980.

Gordon, R. J.: Alternative responses of policy to external supply shocks. *Brookings Papers on Economic Activity*, 1975.

Gramlich, E. M.: Macro policy responses to price shocks. *Brookings Papers on Economic Activity*, 1979.

Hicks, J.: *The crisis in Keynesian economics.* Basil Blackwell, Oxford, 1975.

Hoel, M.: Makroøkonomiske konsekvenser av en sterk økning av råoljeprisen på kort og lang sikt. *Sosialøkonomen*, no. 4, 7–15, 1979.

Hoel, M.: Properties of a short-run production function of the Johansen type with several inputs. Memorandum from the Institute of Economics, University of Oslo (July 16), 1980*a*.

Hoel, M.: A two-sector disequilibrium model with labour immobility. Memorandum from the Institute of Economics, University of Oslo (September), 1980*b*.

Kaldor, N.: Economic growth and the problem of inflation. *Economic 26*, 1959.

Phelps, E. S.: Commodity-supply shock and full-employment monetary policy. *Journal of Money, Credit and Banking, 10*, 1978.

Pindyck, R. S.: Energy price increases and macroeconomic policy. *Energy laboratory working paper no. MIT-EL 79 -061WP*. Massachusetts Institute of Technology, 1979.

Rødseth, A.: Oljepriser og økonomiske kriser. *Sosialøkonomen*, no. 9, 24–30, 1978.

Scitovsky, T.: Market power and inflation. *Economica 45* (August), 221–233, 1978*a*.

Scitovsky, T.: Asymmetries in economics. *Scottish Journal of Political Economy 25* (Nov.) 227–237, 1978*b*.

Solow, R. M.: What to do (macroeconomically) when OPEC comes. Draft paper NBER. Conference on Rational Expectations and Economic Policy, October 1978.

Steigum, E.: Keynesian and classical unemployment in an open economy. *Scandinavian Journal of Economics 82* (2), 147–166, 1980.

THE OPTIMAL PRODUCTION OF AN EXHAUSTIBLE RESOURCE WHEN PRICE IS EXOGENOUS AND STOCHASTIC*

Robert S. Pindyck

Massachusetts Institute of Technology, Cambridge, Massachusetts, USA

Abstract

This paper examines the optimal production of a resource such as oil when its price is determined exogenously (e.g. by a cartel such as OPEC), and is subject to stochastic fluctuations away from an expected growth path. We first examine the dependence of production on extraction cost, and show that the conventional exponential decline curve is indeed optimal if marginal cost is constant with respect to the rate of extraction but is a hyperbolic function of the reserve level. We next show that uncertainty about future price affects the optimal production rate in two ways. First, if marginal cost is a convex (concave) function of the rate of production, stochastic fluctuations in price raise (lower) average cost over time, so that there is an incentive to speed up (slow down) production. Second, the "option" value of the reserve, i.e. the ability to withhold production indefinitely and never incur the cost of extraction, provides an incentive to slow down the rate of production.

I. Introduction

Suppose you owned reserves of an exhaustible resource such as oil. How fast should you produce the resource if its price follows an exogenous growth path, and how should your rate of production be influenced by uncertainty over the future evolution of price?

Hotelling (1931) originally showed that with constant marginal extraction costs, you would produce at maximum capacity or else not at all, depending on whether price net of marginal cost (i.e. "rent") was expected to grow slower or faster than the rate of interest. Thus market clearing would ensure that rent grew at exactly the rate of interest, and producers would be indifferent about their rates of production. But clearly the producers of most resources in competitive markets today are far from indifferent over their production rates, and resource prices (whether in competitive or monopolistic markets)

---

* Research leading to this paper was supported by the National Science Foundation under Grant No. SES-8012667, and this support is gratefully acknowledged. The author also wishes to thank Robert Merton for helpful comments and suggestions.

have usually not grown steadily over time as in the simple version of the Hotelling model.

One reason for this is that marginal production costs for most resources are usually not constant, but instead are likely to vary linearly or nonlinearly with the rate of production, and to depend (inversely) on the level of reserves as well. A second reason is that in most cases resource owners perceive the future price of the resource as uncertain, and even if those owners are risk-neutral, this can (as we will see) lead to a shift in behavior. In this paper we focus on these two issues, but recognizing of course that there are a number of other important factors that may also influence the rate of resource production and the behavior of resource markets.[1]

We will examine the optimal production of a (nondurable) exhaustible resource, e.g. oil, when its price follows an exogenous growth path, and may be subject to stochastic variation around that path. However we will *not* be concerned with the determination of the expected price trajectory, or the reasons for stochastic fluctuations around that trajectory. The reader might like to assume (and it would be reasonable to do so) that the resource price is controlled by a cartel (such as OPEC), and both the expected and realized price trajectories reflect a mixture of rational and (to economists) irrational behavior on the part of the cartel.[2]

We will see in this paper that uncertainty over the future price of the resource can affect the current production rate for two reasons. First, if marginal extraction cost is a nonlinear function of the production rate, stochastic fluctuations in price will lead (on average) to increases or decreases in cost over time, so that cost can be reduced by speeding up or slowing down the rate of depletion. Second, in-ground reserves of a resource can be thought of as an "option" on the future production of the resource; if the future price of the resource turns out to be much higher than the cost of extraction, it may well be desirable to "exercise" the option and produce the resource, but if instead price falls so that production would be unprofitable, the option need never be exercised, and the only loss is the cost of discovering or purchasing the

---

[1] For example, Pindyck (1978*b*) discusses the interrelationship between the rate of production and the rate of exploration and reserve accumulation, Levhari and Pindyck (1981) show how the *durability* of some resources affects their production and market price paths, and Newbery (1980) examines the implications of alternative market structures for production rates. Also, in an earlier paper Pindyck (1980) examined the effect of demand and/or reserve uncertainty on the expected evolution of the competitive market price, but that paper made simplifying assumptions about the characteristics of extraction cost, and also ignores the possibility that producers might withhold production that is currently uneconomical until (and if) price unexpectedly rises.

[2] An earlier paper by this author (1978*a*) examined the optimal price behavior for an exhaustible resource cartel, taking non-cartel supply behavior as exogenous (and not dynamically optimal). For models in which the price and output behavior of both the cartel and the "competitive fringe" are dynamically consistent (i.e. Nash-Cournot models), see Salant (1976) for the case in which all producers face identical costs, and Newbery (1980) for the more general case in which there are differences in costs and/or discount rates.

reserve.[1] But this means that under future price uncertainty the current value of a unit of reserve is *larger* than the current price net of extraction cost, and as we will see, the greater the uncertainty the greater is the incentive to *hold back production*, and keep the option.

In the next section we set forth a simple deterministic model of optimal production in which price is assumed to grow exponentially at a rate less than the rate of interest, and production costs may be a general function of the rate of production and the level of reserves. The solution of that model is straightforward, but it is useful to examine the characteristics of the production trajectory under different assumptions about the cost function and the rate of growth of price. In Section III we introduce uncertainty by letting the price follow a stochastic process so that its future values are lognormally distributed around the expected growth path, and its variance grows linearly with the time horizon. We then solve the resulting stochastic optimization problem, but first ignoring the "option" value of the reserve, i.e. by calculating the expected value of the reserve based on the assumption that it *is* eventually extracted. This will enable us to examine the relationship between price uncertainty, the characteristics of extraction cost, and the rate of production. We will see that price uncertainty leads to faster (slower) production if marginal cost is a convex (concave) function of the rate of production. We then consider the value of an in-ground unit of reserves as an "option" on possible future production, and show that even if producers are risk neutral this implies a *slowing down* of production if there is price uncertainty. Finally we summarize our results and offer some concluding remarks in Section IV.

II. Optimal Production under Certainty

We assume that our resource producer begins with a known reserve level R_0, and has a *total* cost of extraction $C(q, R)$, with $C_q > 0$, $C_{qq} \geqslant 0$, $C_R \leqslant 0$, $C_{qR} \leqslant 0$, $C_{qqR} \leqslant 0$, $C_{qRR} \geqslant 0$, and $C(0, R) = 0$. (For now we will assume that the inequalities hold for all of these partial derivatives; shortly we will consider special cases where some of the partials are equal to zero.) We also assume that the producer knows that the price of the resource p will grow at the rate α, where $0 \leqslant \alpha < r$, and r is the rate of interest.[2] The producer's problem, then, is:

$$\max_{q(t)} \int_0^\infty [p(t)\, q(t) - C(q, R)]\, e^{-rt}\, dt \tag{1}$$

such that

$$\dot{R} = -q, \ R(0) = R_0 \tag{2}$$

The option value of an in-ground reserve is discussed in a recent paper by Tourinho (1979), who uses the standard Black-Scholes option pricing model to show how the value of the reserve grows as the degree of future price uncertainty grows.

[2] Whether price is determined in a quasi-competitive market or is set by a cartel, we would expect it to grow at a rate less than r as long as extraction costs are positive.

$$\dot{p} = \alpha p,\ p(0) = p_0 \tag{3}$$

and $R(t),\ q(t) \geqslant 0$.

This is a straightforward optimal control problem. Define the Hamiltonian H as usual, and maximize with respect to q to get the discounted rent, or shadow price of a unit of reserves:

$$\lambda = (p - C_q)e^{-rt} \tag{4}$$

Now differentiate eqn. (4) with respect to time to get an expression for $\dot{\lambda}$, substitute (2) and (3) for $\dot{p}$ and $\dot{R}$, substitute $\dot{\lambda} = -\partial H/\partial R = C_R e^{-rt}$, and re-arrange to obtain the equation that describes the dynamics of production:

$$\dot{q} = -\frac{1}{C_{qq}}[(r - \alpha)p - rC_q - C_{qR}q + C_R] \tag{5}$$

The optimal production and reserve trajectories are thus determined from the simultaneous solution of the three differential equations (2), (3), and (5), together with the two initial conditions for $R(0)$ and $p(0)$, and one terminal condition. Since at the terminal time T, $H(T) = 0$, the terminal condition will depend on the cost function $C(q, R)$ and the rate of growth of price. If $C_R = 0$ or $C_R < 0$ but $C_{qR} \to a < \infty$ as $R \to 0$, the condition is $q(T) = 0$, $R(T) = (>)0$ if $p(T) - C_q(T) > (=)0$. This condition may also apply if $C_{qR} \to \infty$ as $R \to 0$, but if price grows fast enough, $q(t)$, $R(t)$ and $p(t) - C_q$ will all approach zero asymptotically. Finally, note that the assumption that $\alpha < r$ is essential; if $\alpha \geqslant r$ there is no incentive to produce at all.

The behavior of production is easiest to understand by examining some special cases. We begin with the case in which marginal cost is constant with respect to the rate of production, i.e. $C_{qq} = 0$, but C_R, $C_{qR} < 0$. This implies a singular solution, so that eqn. (5) no longer holds. Instead, maximization of the Hamiltonian gives:

$$q(t) = \begin{cases} \bar{q}_{\max} & \text{if } (p - C_q) > \lambda e^{rt} \\ q^*(t) & \text{if } (p - C_q) = \lambda e^{rt} \\ 0 & \text{if } (p - C_q) < \lambda e^{rt} \end{cases} \tag{6}$$

If marginal cost did not depend on reserves, production would be set at either $\bar{q}_{\max}$ or 0 depending on α, but with $C_{qR} < 0$, the interior solution $q^*(t)$ may apply for at least part of the time. We can determine the interior solution by differentiating the condition $(p - C_q) = \lambda e^{rt}$ with respect to time and rearranging:

$$q^*(t) = \frac{1}{C_{qR}}[(r - \alpha)p - rC_q + C_R] \tag{7}$$

Note that this implies that production is non-zero when $r(p - C_q) < \alpha p - C_R$, i.e. the marginal profit from producing one unit is less than the capitalized value of future gains plus cost savings on *all* units currently extracted if the one unit were to be left in the ground.

Generally production will begin at either 0 or $\bar{q}_{\text{max}}$ and stay there until p and R are such that the interior condition holds, and then production will follow $q^*(t)$ for the remainder of the time. We can see this using as an example $C(q, R) = mq/R$. Now eqn. (7) implies that $p(t) = mr/(r - \alpha) R^3$. Differentiating this with respect to time and substituting αp for $\dot{p}$ gives an expression for $q^*(t)$:

$$q^*(t) = 3\alpha R = 3\alpha \left[\frac{mr}{(r - \alpha) p} \right]^{1/3} \tag{8}$$

so that when the interior solution applies, production is proportional to the reserve level. In fact this gives the familiar exponential decline curve for production, with the decline rate equal to 3α, i.e. $q^*(t) = 3\alpha R_0 e^{-3\alpha(t - t_0)}$. Note that if $\alpha = 0$ the interior solution never applies, and $q = \bar{q}_{\text{max}}$ throughout.

The optimal production and reserve trajectories are shown in Figure 1 for the situation where $q = \bar{q}_{\text{max}}$ initially, and in Fig. 2 for the situation where $q = 0$ initially. In Fig. 1, R_0 and p_0 are such that $R_0 > [mr/(r - \alpha) p_0]^{1/3}$, so that $q = \bar{q}_{\text{max}}$ until R falls and p rises to the point where $R = [mr/(r - \alpha) p]^{1/3}$, after which q and R follow the interior solution given by eqn. (8). Note from Fig. 1 that as α becomes smaller, q remains at $\bar{q}_{\text{max}}$ longer, and the switch to the interior solution occurs at lower values of q and R.

In Fig. 2, R_0 is small relative to p_0 and α, so that $q = 0$ until p rises to the point where $R_0 = [mr/(r - \alpha) p]^{1/3}$, after which q and R again follow the interior solution. Note that if α is made larger, q remains at 0 longer, and the interior solution implies higher values of q and R. Letting T_2 be the time at which q switches from 0 to $q^*(t) > 0$, note that $T_2 \to \infty$ as $\alpha \to r$. As one would expect, the higher the expected rate of price increase, the longer a resource owner would be willing to hold off commencing production. But as long as $\alpha < r$ and $C_{qR} < 0$, the owner will eventually start producing.

We observe, then, that even if marginal cost is constant with respect to the rate of output, dependence on the reserve level will imply an interior solution for the optimal production rate that will hold for at least part of the time. Furthermore, if marginal cost varies hyperbolically with the reserve level (which is at least approximately the case for many oil and gas reserves), the interior solution is the conventional exponential decline curve for production.[1]

Now consider the alternative special case where cost is independent of reserves, i.e. $C_R = C_{qR} = 0$, but $C_{qq} > 0$. In this case an interior solution always applies, and eqn. (5) simplifies to:

$$\dot{q} = -\frac{1}{C_{qq}} [(r - \alpha) p - r C_q] \tag{9}$$

---

[1] The exponential decline curve is used widely by petroleum engineers as a basis for production planning, and it is interesting that it has an economic justification if the cost function is hyperbolic in reserves. For a discussion of the exponential decline curve and its use as a rule of thumb guide, see McCray (1975).

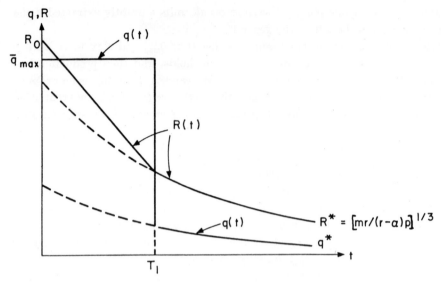

Fig. 1. Production when $C_{qq} = 0$ and R_0 is large relative to P_0 and α.

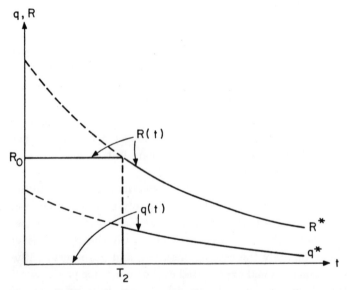

Fig. 2. Production when $C_{qq} = 0$ and R_0 is small relative to P_0 and α.

The characteristics of the solution in this case will depend on both α and the shape of the marginal cost curve, and in particular the sign of C_{qqq}. Suppose $\alpha = 0$. Then clearly $\dot{q} < 0$ to exhaustion. Furthermore, if $C_{qqq} \geqslant 0$, $\ddot{q} < 0$ throughout as well. If $C_{qqq} < 0$ and is sufficiently large in magnitude, $\dot{q}$ will be positive at first and later will turn and remain negative until exhaustion. These possibilities are shown as trajectories A and B in Fig. 3.

q(t)

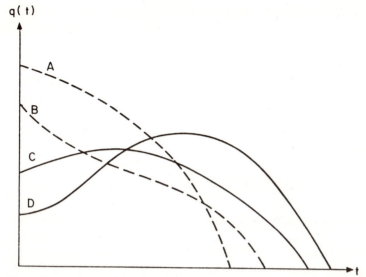

Fig. 3. Production trajectories when $C_R = 0$.

Now suppose $\alpha > 0$. In this case the sign of $\dot{q}$ is not clear, at least during the initial period of production. With $C_R = 0$, discounted rent λ must be constant over time, so that if price is growing fast enough, production may be initially increasing. In such a case, $\dot{q}$ will change sign (q must fall to zero as $R \to 0$ since $H(T) = 0$), and can change sign only once. To see this, note that

$$\ddot{q} = -\frac{1}{C_{qq}}[r^2 \lambda_0 e^{rt} - \alpha^2 p_0 e^{\alpha t}] \tag{10}$$

so that $\ddot{q}$ is negative always, or else $\ddot{q}$ is positive initially, and turns negative later. Thus if α is large, optimal production can also follow trajectories C or D in Fig. 3.

III. Production when Price is Stochastic

To introduce uncertainty over future values of price, we assume that price fluctuates from its expected growth path according to a stochastic process with independent increments. In particular, we replace eqn. (3) with

$$dp = \alpha p\,dt + \sigma p\,dz = \alpha p\,dt + \sigma p \varepsilon(t)\sqrt{dt} \tag{11}$$

where $\varepsilon(t)$ is a serially uncorrelated normal random variable with zero mean and unit variance (i.e. $z(t)$ is a Wiener process).[1] Equation (11) implies that the

---

[1] Eqn. (11) is the limiting form as $h \to 0$ of the discrete-time difference equation $p(t+h) - p(t) = \alpha p(t)h + \sigma p(t)\varepsilon(t)\sqrt{h}$ and $E[dp/p] = \alpha dt$, and Var $[dp/p] = \sigma^2 dt$. Note that $p(t)$ is log-normally distributed, with $E_0[\log (p(t)/p(0))] = (\alpha - \frac{1}{2}\sigma^2)t$, and Var $[\log (p(t)/p(0))] = \sigma^2 t$. For an introduction to stochastic differential equations of the form of (11), see Cox & Miller (1965).

current price is known exactly, that uncertainty about future price grows with the time horizon, and that fluctuations in price occur continuously over time. (Note that continuity does not rule out rapid changes in price; over any finite time period any change in price of finite size is possible. On the other hand, whether the past behavior of oil prices can be best represented by a continuous process or a "jump" process is an empirical question that has not been resolved.)

We assume that producers are risk neutral, so that the dynamic optimization problem is now:

$$\max_{q(t)} E_0 \int_0^\infty [p(t)\,q(t) - C(q, R)]\,e^{-rt}dt = E_0 \int_0^\infty \prod_d(t)\,dt \tag{12}$$

subject to the ordinary differential equation (2), the stochastic differential equation (11), and $R(t)$, $q(t) \geqslant 0$.

Our approach is to first solve this problem under the assumption that p_0 and α are such that $q(0) > 0$, and that $q(t) > 0$ over the entire planning horizon (i.e. up to the point where exhaustion occurs or $C_q = p$). This can be done using stochastic dynamic programming, and although it ignores the possibility of withholding all production (perhaps indefinitely), it will allow us to determine how the effect of price uncertainty on current production depends on the characteristics of cost. Afterwards we will consider the ability of the producer to withhold production (of perhaps currently uneconomical reserves), while maintaining the option of producing in the future should price unexpectedly rise rapidly. As we will see, this leads to a quite different effect of price uncertainty.

We begin, then, by looking for an interior solution to the optimization problem set forth above. To do this, define the optimal value function:

$$J = J(R, p, t) = \max_{q(\tau)} E_t \int_t^\infty \prod_d(\tau)\,d\tau \tag{13}$$

Since J is a function of the stochastic process p, the fundamental equation of optimality is:[1]

$$0 = \max_{q(t)} \left\{ \prod_d(t) + (1/dt)\,E_t\,dJ \right\}$$
$$= \max_{q(t)} \left\{ \prod_d(t) + J_t - qJ_R + \alpha p J_p + \frac{1}{2}\sigma^2 p^2 J_{pp} \right\} \tag{14}$$

Maximizing with respect to q gives:

$$\partial \prod_d / \partial q = J_R \tag{15}$$

---

[1] We use the notation $J_R = \partial J/\partial R$, etc. $(1/dt)\,E_t\,d(\ )$ is Ito's differential generator. For a discussion of the techniques used in this paper, and in particular the use of Ito's Lemma, see Kushner (1967), Merton (1971), or Chow (1979). Also, the approach here follows closely that used in this author's (1980) earlier paper on this subject.

i.e. the usual result that the shadow price of the resource should always equal the incremental profit that could be obtained by selling an additional unit. Now differentiate eqn. (14) with respect to R:

$$\frac{\partial \Pi_d}{\partial R} + J_{Rt} - qJ_{RR} + \alpha p J_{Rp} + \frac{1}{2}\sigma^2 p^2 J_{Rpp} = 0, \tag{16}$$

and by Ito's Lemma note that this can be re-written as:

$$\partial \Pi_d / \partial R + (1/dt) E_t d(J_R) = 0. \tag{17}$$

To eliminate J from the problem, apply the operator $(1/dt) E_t d(\ )$ to both sides of (15), and combine the resulting equation with (17) to yield:

$$(1/dt) E_t d(\partial \Pi_d / \partial q) = -\partial \Pi_d / \partial R. \tag{18}$$

Eqn. (18) is just a stochastic version of the well-known Euler equation from the calculus of variations. In its integral form it says that the marginal profit from selling 1 unit of reserves should just equal the expected sum of all discounted future increases in profit that would result from holding the unit in the ground.

Our objective is to obtain an equation analogous to eqn. (5) to explain the *expected* dynamics of production. To do this, substitute $\partial \Pi_d / \partial q = [p(t) - C_q]e^{-rt}$ and $\partial \Pi_d / \partial R = -C_R e^{-rt}$ into eqn. (18):

$$-r[p(t) - C_q] + (1/dt) E_t dp - (1/dt) E_t dC_q = C_R. \tag{19}$$

Now note that $E_t dp = \alpha p \, dt$, and expand the stochastic differential dC_q using Ito's Lemma:

$$dC_q = C_{qq} dq + C_{qR} dR + \tfrac{1}{2} C_{qqq}(dq)^2 + o(t) \tag{20}$$

where $o(t)$ represents terms that vanish as $dt \to 0$. Along an optimal trajectory $q = q^*(R, p)$, so that

$$E_t(dq)^2 = \sigma^2 p^2 q_p^2 dt + o(t) \tag{21}$$

where q_p is the (unknown) response of optimal production to a change in price. Now substituting eqns. (20) and (21) into (19) and rearranging, we obtain the equation, analogous to eqn. (5), that describes the expected dynamics of production:

$$\frac{1}{dt} E_t dq = -\frac{1}{C_{qq}}\left[(r - \alpha)p - rC_q - C_{qR}q + C_R + \frac{1}{2}\sigma^2 p^2 q_p^2 C_{qqq} \right] \tag{22}$$

Eqn. (22) tells us that the expected rate of change of production differs from the certainty case when marginal cost C_q is a nonlinear function of the rate of production. In particular, we see that when production is falling, price uncertainty causes it to fall faster (slower), so that production begins at a higher (lower) level, when marginal cost is a convex (concave) function of q.

This deviation from the certainty case is easily understood by recognizing that stochastic variations in price imply changes in expected future marginal costs if $C_{qqq} \neq 0$. To see this, suppose $C_{qqq} > 0$, and random increases and decreases in price occur that balance out, leaving price unchanged on average. Clearly such fluctuations will have the net effect of raising marginal cost over time, since corresponding increases in optimal production will raise marginal cost more than corresponding decreases will lower it. This in turn implies an incentive to speed up production, and thereby reduce these expected increases in cost over time. If, on the other hand, $C_{qqq} < 0$, just the opposite holds, and there is an incentive to slow down production.

We see then that if marginal cost increases nonlinearly, price uncertainty will lead to changes in the expected rate of production. Does this mean that if marginal cost is constant or rises linearly with the rate of production, price uncertainty should have no effect on expected production? The answer is no, as we see if we remember that producers need not produce at all. This means that if current or *expected* price is below extraction cost for any marginal unit of the resource, the owner can keep the unit in the ground indefinitely but maintain the option of extracting it at some future time in the event that there is a sufficient (random) increase in price. Alternatively, suppose that production of a unit would be profitable, but just barely so. Then with future price sufficiently uncertain there is an incentive to keep the unit in the ground, since if price were to fall the only loss would be the small (unrealized) profit, whereas if price were to rise, extraction could then yield a relatively large profit.

This means that uncertainty over future price creates an incentive to *slow down* production. To see this, assume that $C_q(0, R) > 0$, and suppose for simplicity that $C_{qqq} = C_R = 0$. Now consider the present value of a marginal unit of reserves that can be extracted at some time t. *Under certainty* that value is

$$V = \max \{0, (p - C_q) e^{-rt}\} \tag{23}$$

i.e. extraction of the unit would occur only if $p(t) > C_q$. Further, it is easy to see from (23) that the value is constant over time, i.e. $dV/dt = 0$.

Now suppose future price is uncertain. The present value of the marginal unit is then just the expected value of the right-hand side of eqn. (23). Now consider the expected rate of change of that value. Since the right-hand side of (23) is a convex function of p, we have by Jensen's inequality:

$$(1/dt) E_0 dV = (1/dt) E_0 \max \{0, d[(p - C_q) e^{-rt}\}$$
$$> (1/dt) \max \{0, E_0 d[(p - C_q) e^{-rt}]\} = 0. \tag{24}$$

Thus under uncertainty it is preferable *not* to extract the marginal unit that under certainty would have been extracted at time t, since that unit is now expected to rise in value over time. This means that given any particular current price (and given any expected rate of price increase α), production should be lower the greater the uncertainty over future price.

Uncertainty over future price therefore affects current production in two different ways. First, if marginal cost is a nonlinear function of the rate of production, future price uncertainty creates a cost-based incentive to alter the current rate of production—speeding it up if the marginal cost function is convex, and slowing it down if it is concave. Second, whatever the characteristics of marginal cost, the fact that any in-ground unit has value as an option to extract—or not extract—in the future implies that future price uncertainty will cause a slowing down of production. Of course if the marginal cost function is convex, the net effect is ambiguous.

IV. Concluding Remarks

Models of resource production often contain severe simplifying assumptions about the characteristics of cost and about knowledge of future price, and we have seen in this paper that such assumptions may lead to highly misleading results. For example, for most resources, and certainly for oil and gas, extraction cost is in fact usually not constant with respect to the reserve level and the rate of production as often assumed, and this means that resource owners should be far from indifferent about their rates of production. In particular, if marginal cost can be roughly characterized as a hyperbolic function of the reserve level, then the conventional exponential decline curve often used by petroleum engineers will apply.

It is also a fact that there is considerable uncertainty about the future prices of most resources. This is particularly true today for oil and other energy resources, where price determination by a politically unstable cartel makes market evolution highly unpredictable. We have seen that even if resource owners are risk neutral, uncertainty over future price will alter their current rates of production, in a way that depends on the characteristics of marginal cost.

Of course we introduced uncertainty in this paper by letting price follow a continuous stochastic process, and as we mentioned earlier, it is an empirical question as to whether past price behavior is best described by a continuous process or a "jump" (discontinuous) process. It is unlikely, however, that our qualitative results will depend strongly on the assumption of continuity, and roughly the same results should obtain if a "jump" process were used to describe price, as long as the jumps are reasonably frequent. We leave it to others to demonstrate that this is indeed the case—or to show that it is not. Our assertion that the continuity assumption probably does not matter much is an intuitive one based on analogous problems in finance theory; as Merton (1976) has shown, option pricing formulas for continuous stock returns provide a good approximation even if actual returns are discontinuous, as long as the fluctuations in those returns occur frequently.

Finally, it should be stressed that in this paper price was exogenous, and we did not consider market equilibrium. This is reasonable in the case of oil, where price is now controlled (rationally or irrationally) by a cartel. The earlier paper by this author (1980) examined the effects of future demand and reserve uncertainty on the evolution of competitive market price, but made the assumption that marginal cost is constant with respect to the rate of production, and ignored the value of in-ground reserves as an option (that need not be exercised) on future production. In so doing it found that with constant marginal cost, under demand and/or reserve uncertainty the expected rate of change of rent, i.e. the competitive market price net of marginal cost, would still equal the rate of interest. It can be seen from eqn. (24), however, that because of the option value of in-ground reserves, this is in fact not the case, and in expected value terms rent should rise at *less* than the rate of interest, since otherwise the expected present value of in-ground reserves would rise over time.

References

1. Chow, Gregory C.: Optimal control of stochastic differential equation systems. *Journal of Economic Dynamics and Control 1*, April 1979.

2. Cox, D. R. & Miller, H. D.: *The theory of stochastic processes*. Chapman and Hall, London, 1965.

3. Hotelling, Harold: The economics of exhaustible resources. *Journal of Political Economy 39*, 137–175, April 1931.

4. Kushner, Harold J.: *Stochastic stability and control*. Academic Press, New York, 1967.

5. Levhari, David & Pindyck, Robert S.: The pricing of durable exhaustible resources. *Quarterly Journal of Economics*, August 1981.

6. McCray, Arthur W.: *Petroleum evaluations and economic decisions*. Prentice-Hall, Englewood Cliffs, 1975.

7. Merton, Robert C.: Optimum consumption and portfolio rules in a continuous-time model. *Journal of Economic Theory 3*, 373–413, December 1971.

8. Merton, Robert C.: Option pricing when underlying stock returns are discontinuous. *Journal of Financial Economics 3*, 125–144, 1976.

9. Newbery, David M. G.: Oil prices, cartels, and a solution to dynamic consistency. Unpublished working paper, Churchill College, Cambridge, England, May 1980.

10. Pindyck, Robert S.: Gains to producers from the cartelization of exhaustible resources. *Review of Economics and Statistics 60* (2), 238–251, May 1978*a*.

11. Pindyck, Robert S.: The optimal exploration and production of nonrenewable resources. *Journal of Political Economy 86* (5), 841–862, October 1978*b*.

12. Pindyck, Robert S.: Uncertainty and exhaustible resource markets. *Journal of Political Economy 88* (6), 1203–1225, December 1980.

13. Salant, Stephen W.: Exhaustible resources and industrial structure: A Nash-Cournot approach to the world oil markets. *Journal of Political Economy 84* (5), 1079–1094, October 1976.

14. Tourinho, Octavio A. F.: The option value of reserves of natural resources. Working Paper No. 94, Graduate School of Business, University of California at Berkeley, September 1979.

RESOURCE PRICING AND TECHNOLOGICAL INNOVATIONS UNDER OLIGOPOLY: A THEORETICAL EXPLORATION*

Partha Dasgupta

London School of Economics, London, England

Abstract

This article presents a non-technical survey of some recent theoretical analyses of the pricing of exhaustible natural resources and the incentives for developing resource substitutes under oligopoly, and looks in perspective at what appear to be the central considerations. Among the main conclusions that are presented are those that bear on the pace of R & D activity, the phenomenon of sleeping patents, oligopolistic pricing before and after the resource-substituting invention, the timing of innovation, and the transition from one resource base to another. It concludes with an analysis of a national R & D policy for a resource-importing country.

I. Introduction

In this paper I present certain considerations that appear to be unavoidable in any analysis of the pricing of exhaustible natural resources under oligopolistic market conditions. The treatment will be theoretical. Nevertheless, the motivation behind the specific modelling that I appeal to here arises from a need to understand the mode of pricing of exhaustible *energy* resources, such as fossil fuels.[1] In particular, I shall attempt to relate oligopolistic modes of pricing with the incentives that exist for the introduction of substitute sources of energy, such as shale oil or, at some steps removed, controlled nuclear fusion or clean fast-breeders—backstop technologies, as some would call them. Such models would seem to be of interest on their own, for they enable one to study the interplay between rent and production based capital theory. But at the more practical level, considerations such as the ones that I shall discuss here are of importance if we were, for example, to ask whether OPEC's pricing policy is in its *own* long term interest or whether, as some have remarked, it has been setting the price of crude oil at too high a level, in the sense that it has

---

* I am most grateful to Richard Gilbert, Geoffrey Heal and Joseph Stiglitz for illuminating discussions over the past several years on the problems discussed in this essay. For his penetrating comments on an earlier version of this essay I am indebted to Dilip Mookherjee.
[1] Formal derivations of most of the results that I shall discuss here are provided in Dasgupta & Stiglitz (1976, 1980, 1981) and Dasgupta, Gilbert & Stiglitz (1980, 1981).

provided excessive incentives to others for the invention of substitute energy sources, whose existence will undercut OPEC's monopoly power. Now it is clear enough in advance that such questions cannot be answered in the absence of an explicit reference to some construct that builds on OPEC's intertemporal oligopolistic behaviour. To be sure, this was recognized in the early simulation studies of Cremer & Weitzman (1976) and Pindyck (1978). The problem with these works, as with most other simulation studies, was that they did not provide an account of why the answers emerged the way they did.[1] Moreover, they explicitly avoided mention of substitute sources of energy, and thereby were incapable of addressing themselves to the central question I ask here. Furthermore, one gets the unmistakable impression that the model specifications were chosen with a view to an *ex post* prediction of the magnitude of the 1973 oil price rise.

Any analysis of oligopolistic pricing of exhaustible natural resources invites us to explore dynamic games (i.e. games in extensive form) with all their attendant difficulties. To be sure, the presence of such resources imposes a strong structure on the characteristics of intertemporal game equilibria.[2] This, as we shall see, allows us to locate some of the salient features of these equilibria relatively easily. Nevertheless, it is clear enough that detailed features of the price trajectory—and hence the pace of extraction—of an exhaustible natural resource in oligopolistic markets depend, among other things, on the strategies available to the agents, their goals, the conjectures they hold, and on whether or not binding agreements can be reached among some or all of the agents concerned. These are familiar matters, but they are worth reiterating if only to emphasize the obvious truth that economists are unlikely to reach a consensus about how one might best model oligopolistic behaviour in the abstract.

In this paper I shall review a very simple construction—almost an example —which will enable me to discuss resource depletion and resource-substituting technological innovation under oligopolistic conditions. In order to concentrate on essential matters I shall be thinking of a single-grade stock, and I shall consider a well-defined backstop technology that enables one to produce a perfect substitute at constant unit cost. To have an interesting problem one supposes that the unit cost of extraction is less than the cost of production of a unit of the substitute. Thus the model captures the essence of a tension that is provided by the *simultaneous* presence of a *finite* resource base which is easy to exploit and an *infinite* resource base which is costly to exploit.[3] But

---

[1] For a theoretical analysis of oligopolistic models similar to that of Cremer & Weitzman (1976), see Gilbert (1978) and Dasgupta & Heal (1979), chapter 11, section 7.

[2] The classical Hotelling Rule, which characterizes the price trajectory along an intertemporal competitive equilibrium, is an illustration of this.

[3] One must not take the notion of an *infinite* resource base—a backstop technology—too literally. The tension I am talking about is really provided by the simultaneous presence of a "small" deposit which is "cheap" to extract and a "large" deposit that is expensive to tap. The presence of many deposits of varying qualities and sizes can be easily accommodated. See Dasgupta & Heal (1979), chapter 6 and 11.

this situation, the subject of Section III, is merely a necessary prelude to the central case I wish to discuss; namely the case where the backstop technology is not currently at hand, and where firms (or governments) are capable of investing in research and development (R & D) to make the new technology *viable*.

Since it is a point of considerable importance it is as well to emphasize the fact that it is the *size* of the existing resource *stock*, and *not* the prevailing resource *price*, which determines the incentives that firms (or governments) at any date have for undertaking R & D expenditure designed to make a backstop technology viable. To see this most clearly, suppose for example, that the exhaustible resource is under the control of a monopolist (e.g. a cartel) and R & D competition is among potential suppliers. It should now be clear why, somewhat paradoxically, the persistence of high resource price does not make for a favourable economic climate for R & D undertaking by rival firms. For, if the resource price is "high" over a period, sales over this period are "low". But in this case the stock remaining at the end of the period will be "large". The resource cartel will therefore be able to undercut owners of the backstop technology for just that much longer, delaying the date at which *innovation* can profitably be made. This is important to bear in mind. If OPEC can raise its price when it is profitable to do so, it can presumably lower its price if that is the profitable move. But a recognition of this possibility dampens the enthusiasm with which R & D activity is undertaken by others.

There are two broad categories of cases that need to be distinguished: the incentives that potential *suppliers* have for developing a backstop—the case just considered—and the incentives that resource *users* have in having it developed. The former are concerned only with the profitability of sales after development has been completed, and this depends critically on reserves at the date of invention. But the latter enjoy, in addition, benefits before completion date. This is because announcement of an invention date influences the cartel's pricing policy. In Sections IV and V, I shall analyze the incentives to each such type in turn by means of two examples.

In what follows I shall, for simplicity of exposition, introduce complications sequentially. In Section II, I discuss *competitive* extraction and suggest that *uncertain* property rights—in this case the threat of expropriation by host countries—result in foreign extractors attaching a risk-premium on the rate of return they use for extraction decisions, and that this results in *excessive* depletion, and therefore, in too *low* a resource price.

In Section III I suppose that in addition to the exhaustible resource there is a viable backstop technology. The industrial organization that I emphasize here is a duopoly structure: that is, a single resource owner and a single manufacturer of the substitute product.[1] The idea is to analyse an intertemporal

---

[1] The motivation is to see the effect of patent rights on the new technology.

duopoly equilibrium; an analysis which is a necessary prelude to a discussion of a central problem of this paper; namely the incentives for developing a backstop technology under oligopoly. This is discussed in Section IV. It is assumed in Section IV that the backstop technology is currently unavailable, but that the date at which it can be made viable by a firm is a declining function of the expenditure it undertakes. Thus, the date of invention is endogenous to the construct. I shall assume that there is competition in R & D and that the first firm to make the technological invention is awarded the patent. The model is stylized: it supposes that the winner takes all. Thus, the post *invention* era is characterized by a duopoly structure (unless the resource owner is the winner of the patent race!). The analysis of Section III can therefore be regarded as an account of the post-invention era. The conclusions that seem to me to be of particular interest are:

(1) The price at which a resource is marketed in initial years under a duopoly market structure may well be *higher* than under *pure monopoly*. That is, duopoly may encourage more resource conservation than pure monopoly; Section III. (That there is excessive resource conservation under pure monopoly, and delay in the innovation date, as compared to the perfectly competitive case is, of course, well known. See Propositions 2 and 3 below.)

(2) The resource owner has a greater incentive to own the patent then anyone else. What this suggests is that we should expect resource controlling oligopolists to be in the vanguard of technological change which will usher in the new era. That is, it is not simply because of a greater expertise in these matters that oil corporations are among the most active participants in the race for new energy sources.[1]

(3) Competition in R & D may result in the phenomenon of "sleeping patents" in the post-invention era, where, for a period the patent holder does not innovate. That is, the date of *invention* (i.e. the date at which the backstop technology is made viable) need not coincide with the date of *innovation* (i.e. the date at which the new technology is put to use). What is more, the dynamic equilibrium sustaining the sleeping patent here is a *perfect* equilibrium, in the sense that the future patent holder not only supposes that he will hold a sleeping patent when making his R & D decision, but in fact subsequent to invention finds it profitable to withhold the product from the market for a period envisaged earlier.

---

[1] If patents do not provide an adequate protection, ensuring that the invention is kept a secret is an alternative open to the resource cartel. Suppression of the invention enables the monopolist to earn greater rents from its remaining reserves. It is this fact that was dramatized in the recent M. G. M. film *The Formula*. The formula in question pertained to a particularly cheap method of coal liquifaction. The film depicted the draconian measures that an oil baron undertook in order to obtain the formula. Having failed in his attempts, he made a substantial payment to the person in possession of it to have it suppressed for a stipulated period.

(4) The transition from the exhaustible resource base to the inexhaustible one takes place gradually under duopoly. Unlike the polar market structures of perfect competition and pure monopoly, resource exhaustion does *not* precede technological innovation. There is a period during which the duopolists share the market. That is, the resource owner during this period does not find it profitable to undercut the manufacturer of the substitute product. During this period the share of the market controlled by the manufacturer increases as the resource stock dwindles to zero.[1]

As I have mentioned earlier, I shall, in Section IV, be postulating that the natural exhaustible resource has a single owner. In fact it is possible to show that under a wide class of market structures the incentive for undertaking R & D expenditure towards the discovery of substitute sources is non-increasing as a function of the existing resource stock,[2] a point I raised earlier. From this one may conclude with:

(5) Other things being the same, a resource-substituting invention is likely to occur earlier in an economy possessing a low resource base as compared to one that is resource rich. While seemingly banal, this conclusion gives a theoretical underpinning to the thesis that by the 18th century the incentives that England had for developing coal based industry was greater than any of her competitors because of a shortage of timber.[3]

In Section V I consider the kinds of consideration that are involved in the choice of a *national* R & D programme when the country in question relies on a foreign cartel for the supply of an exhaustible natural resource. To capture the essentials of the problem it is appropriate to suppose that the cartel's market is restricted to this country (and that the government of the importing nation proposes to make the backstop technology freely available to manufacturers). My idea will be to draw out the social cost-benefit rule that ought to be used in reaching a national R & D policy in such circumstances. Throughout I assume that the cartel is motivated solely by the present discounted

---

[1] For an illuminating discussion of the "transition" problem and the world's prospects for an inexhaustible base for energy, see Koopmans (1980).

[2] See Dasgupta, Gilbert & Stiglitz (1980).

[3] Brinley Thomas has recently made a forceful presentation of this thesis. "In the second half (of the 18th century) Britain experienced an energy crisis .... After the middle of the 18th century there was a growing general shortage of timber and especially charcoal, whereas coal was relatively plentiful .... The sectoral imbalance could only be resolved by switching the energy base from wood to coal ... but this could not be done until a fundamental innovation had been made .... At the beginning of the 1780's Britain was in trouble, a special kind of trouble which no other country faced at that time. The First British Empire had collapsed and she was in a severe energy crisis, dangerously dependent on wood fuel .... In the previous 25 years as many as 13 inventors had attempted to overcome the technical difficulties. It was not until 1784 that the problem of refining pig iron with coal or coke was finally solved by Henry Cort's process ... the industrial revolution occurred in Britain at the end of the 18th century not because Britain was "well-endowed" in various respects, but because she was "ill-endowed" in one fundamental respect—she was running out of energy and had to do something about it. France had no such problem." Thomas (1980), pp. 2, 11, 12.

value of its rents, and that the importing nation knows this to be the case Indeed, it is then possible to infer:

(6) If the cartel's stock is very large the social return on current R & D expenditure is so low that it is not worthwile undertaking R & D activity. But the social return increases as the cartel depletes its stock, so that at some future date it will typically be worth initiating an R & D programme.

It is also possible to show:

(7) If the cartel's stock is not too large it may well be desirable for the government to embark on an R & D programme whose completion date is expected to occur *before* the cartel exhausts its entire stock. But so long as stocks last the cartel will be able to undercut supply from the backstop technology. Thus technological innovation, in this case, will occur some time after the date of invention of the backstop. It is, nevertheless, desirable for the government to embark on such a "crash" programme, precisely because this is a *credible* means of forcing the cartel to set its current price lower than it otherwise would.

In formulating the government's planning problem we shall also note:

(8) that an exclusive reliance on *marginal* social cost-benefit analysis of R & D programmes is likely to yield sub-optimal policies. Indeed, there are serious non-convexities in the structure of such R & D problems that require an appeal to *global* (social) cost-benefit analysis.

It is natural to compare the socially optimal speed of R & D with the speed at which R & D is undertaken under competition. For the model at hand we shall note that, other things remaining the same, the private reward to a patent holder is less than the social benefit that can be generated by government investment in R & D. However, competition among rivals reduces the present value of profits to the patent holder to zero. Thus we note:

(9) Competition in R & D may well result in excessive R & D activity, in the sense that the technological invention may occur earlier than it would have under the optimal R & D strategy.

Finally, in Section V I present a brief commentary on the kinds of extensions that may be made to the basic model discussed in this paper.

II. The Competitive Arbitrage Condition under the Threat of Expropriation

The analysis that follows is strictly partial-equilibrium in nature. We concentrate our attention on a single industry and suppose the capital market to be perfectly competitive with a constant rate of interest $r > 0$. If the resource is an input in production, its market demand is a derived demand. For ease of exposition I shall suppose in what follows that market demand is positive no matter how high the price. This will enable me to avoid discussing some purely technical matters. If Q_t is the output flow of the industry in question at date t, we let $p(Q_t)$ denote the market clearing price, where it is assumed that $p'(Q) < 0$. A shifting demand curve, something I avoid here, can readily be accommodated.

I want now to consider the asset market. To avoid inessential technicalities I suppose throughout that extraction costs associated with the exhaustible natural resource in question are negligible. In this section I suppose that the resource is owned competitively. But I suppose that there are uncertain property rights. To be precise, suppose resource owners face a common risk of expropriation of their stocks. In particular, let λ_t be the probability rate of expropriation at date t conditional on the entire stock not having been expropriated until then. To make the point I wish to make here in the sharpest manner possible suppose that resource owners expect no compensation if and when the fields are expropriated.[1] Let p_t be the spot price of the resource at t if the stock has not been expropriated by then. Consider a short interval of time $(t, t+\theta)$. If the resource stock is not expropriated during this interval the spot price will be $p_{t+\theta} = p_t + dp_t$. Now, a resource owner has the option of selling a unit of the resource at t and being assured of $(1+r\theta)p_t$ units of account at $t+\theta$. Alternatively he can hold onto this unit. The probability that this unit will be expropriated during $(t, t+\theta)$ is $\lambda_t\theta$. If this occurs then he obtains nothing at the end of the period. Thus the expected value, at $t+\theta$, of a unit of the resource is $(1-\lambda_t\theta)(p_t + dp_t)$. Therefore, if resource owners are risk-neutral— and they will be if there is a complete set of contingent commodity markets —then in dynamic equilibrium we shall have

$$(1-\lambda_t\theta)(p_t + dp_t) = (1+r\theta)p_t. \tag{1}$$

On letting $\theta \to 0$ we can therefore obtain the basic arbitrage condition in continuous time as

$$\dot{p}_t/p_t = r + \lambda_t, \tag{2}$$

where $\dot{p}_t$ is the time derivative of p_t.

Notice that equation (2) is a generalization of the classical Hotelling Rule. In the absence of any threat of expropriation $\lambda_t = 0$, and in this case equation (2) is the Hotelling Rule, which says that the royalty price of an exhaustible natural resource must rise at a percentage rate equal to the rate of interest under competitive conditions. The threat of expropriation results in resource owners using a risk-premium, λ_t. But the point I want to make here is that since the threat of expropriation means that the (spot) resource price rises at a rate in *excess* of r, the price level in initial years is *lower* if there is a threat of expropriation than it would have been were there no such threat. But this in turn

---

[1] The general case is discussed in the context of a different model in Dasgupta & Stiglitz (1981).

means that the risk of expropriation results in *excessive* depletion. We summarize this in the form of

Proposition 1. Suppose that there is a common risk of expropriation of the stocks owned by competitive resource owners, and suppose that no compensation is expected in the event of expropriation. In such an environment resource owners use a risk-premium—equal to the conditional probability rate of expropriation—in assessing the rate of return in holding on to their stock. This means that the threat of expropriation results in excessive depletion in initial years; that is, in initial years the resource is marketed at a price below its efficiency price.

Interest in *Proposition 1* is clearly not merely academic. It is often suggested that the price of Arab crude oil in the decade before 1974 was too low precisely because OPEC left much of the production and pricing decisions to foreign corporations.[1] Admittedly, the oil market, even prior to December 1973, was far from being perfectly competitive. But the general moral that emerges from *Proposition 1* is that the threat of expropriation results in a *bias* towards excessive depletion as compared to an efficient outcome. In what follows I shall ignore the possibility of expropriation and consider the bias towards *excessive conservation*—a bias towards prices in excess of the efficient ones—that arises in oligopolistic markets.

III. The Transition Problem under Alternative Market Structures

III.1. *Introduction*

If we are to obtain an understanding of the relationship between the market price of a resource and the incentives that firms have for undertaking R & D expenditure designed to produce resource-substitutes, we need first to study the market in the *post-invention era*. That is, we need first to study an industry in which both the resource and the backstop technology are available. The reason is that it is the outcome in the post-invention era that determines the reward that can be captured by the firm which develops the backstop technology. Therefore I suppose in this section that the backstop technology is available and that it enables the substitute product to be manufactured at a constant unit cost $C(C>0)$. Let S_0 denote the resource stock at $t=0$, the date at which the invention is made. So as to simplify the exposition I assume away uncertainty. I first look at a competitive industry with a view to locating the efficient outcome. I shall then describe the outcome under pure monopoly.

---

[1] Agricultural economists have long noted an analogue to *Proposition 1* in the context of land degradation in the absence of tenurial rights among peasants. In the case of oil fields, the Rule of Capture, often invoked to cope with the common pool phenomenon, results in excessive drilling. This phenomenon has been much discussed in the literature and is related to the Riparian Doctrine invoked to resolve the uncertain property rights over underground water basins. See, e.g. Khalatbari (1977), Dasgupta & Heal (1979), chapter 12, and Dasgupta (1981).

Finally I study the duopoly model and argue that if reserves at the date of invention are "large", duopoly results in the most *conservative* of these outcomes, in the sense that the resource is marketed at a higher price in initial years under duopoly than even under pure monopoly. I think this is a very suggestive result. For, it seems to me that conventional analysis of oligopolistic markets have suggested that the outcome under oligopoly lies in some sense between outcomes under perfectly competitive and monopolistic industrial structures. If we were to interpret this literally we would wish to claim that the price at which an exhaustible natural resource is marketed is bounded *below* by its competitive price and *above* by its pure monopoly price. But this would be a wrong claim. Just as uncertain property rights may result in the price of a resource being set *below* its competitive price—a result we noted in the previous section—oligopolistic competition may well result in a resource being sold at a price in *excess* of its monopoly price. One can now see why the result is awkward for the analyst. In studying whether OPEC has currently set too high a price for its crude oil even from the point of view of its own interests it will not do to estimate the monopoly price with a view to locating an upper bound for the oligopoly price. In initial years the oligopoly price may well be higher, possibly a good deal higher, than even the monopoly price.[1]

III.2. *The Competitive Case*

Suppose both the resource and the backstop technology are competitively owned. The intertemporal equilibrium outcome for this case is well known.[2] We note it here in the form of

Proposition 2. Along a competitive equilibrium path resource exhaustion precedes the technological innovation. During an initial interval the backstop technology is held in abeyance and the entire market is served by resource depletion. During this interval the resource price rises at a percentage rate equal to the rate of interest, r. The initial price gets so chosen that at the date the resource price reaches the value C (the unit cost of production associated with the backstop technology) the stock is exhausted. From this date the market is served by the backstop technology at the competitive price C; see Fig. 1.

I want to make three observations about *Proposition 2*. First, the transition from the exhaustible resource base to the inexhaustible one is "sharp": the two resource bases are not exploited simultaneously. Considerations of intertemporal efficiency dictate that the cheaper source be exploited first.[3] Second, the price trajectory is continuous: it rises continuously until reaching the long run

---

[1] Hoel (1978), noted this fact in a somewhat different oligopoly structure.

[2] See, for example, Dasgupta & Heal (1979), Chapter 6.

[3] I am appealing to the coincidence of efficient and competitive paths for the model here so as to be able to talk of the dictates of intertemporal efficiency in interpreting *Proposition 2*, even though the Proposition characterizes a competitive outcome.

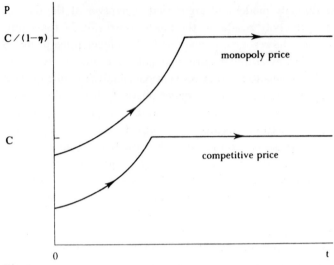

P

C / (1−η)

monopoly price

C

competitive price

0   t

Fig. 1

value of C. Third, and this is intuitively obvious, the larger the initial reserves the lower the initial price and the longer it is until the transition to the new technology is made. It transpires that each of these features is a characteristic of monopoly also, even though the extraction policy of the monopolist is excessively conservative.

III.3. *Pure Monopoly*

I shall now discuss briefly the other polar market structure, one where both the resource and the backstop technology are under the control of a single private owner. So as to make the comparison meaningful I suppose that the monopolist has access to a perfect capital market with a rate of interest r. For obvious reasons I must now suppose that the market demand function has the property that marginal revenue is a declining function of output. Let $\eta(Q) = -Qp'(Q)/p(Q)$ be the inverse of the elasticity of demand. I take it that $\eta(Q) \to \bar{\eta} < 1$ as $Q \to 0$ and that $\eta'(Q) \geqslant 0$. This last is a technical assumption which will enable me to state sharp results. I take it that the monopolist chooses his extraction and production strategy with a view to maximizing the present discounted value of the sum of his net profits. In Section IV I shall need a notation to denote the maximum present value of profits. Let me denote it by $\pi^m(S_0)$; (the superscript m denoting the fact that we are concerned with a monopolist). Now, it is immediate that $d\pi^m(S_0)/dS_0 > 0$. It is also clear enough that in maximizing the present value of his profits, the monopolist will set his marginal revenue equal to the unit cost of production, C, after the transition is complete. Let me denote this long run monopoly price by p^m.[1] Since $p^m > C$,

---

[1] Thus in the long run, after the transition has been completed the monopolist will choose Q in accordance with the rule $p(Q)[1 - \eta(Q)] = C$. In the case where $\eta'(Q) = 0$, we have the familiar fact that $p^m = C/(1 - \eta)$.

we know that under monopoly the long run price of the product exceeds the long run competitive price. It is the period before the transition is complete which requires characterization. This is provided by

Proposition 3. Under pure monopoly resource exhaustion precedes the technological innovation. During an initial interval the monopolist holds the backstop technology in abeyance, and the entire market is served by resource depletion. During this interval the monopolist's resource depletion policy is guided by the rule that marginal revenue rises at a percentage rate equal to the rate of interest. The initial price is so chosen by the monopolist that at the date resource price reaches the long run monopoly price, p^m, the stock is exhausted. From this date the market is served by the backstop technology; see Fig. 1.

It should be noted that each of the three features of the competitive outcome that I highlighted in the previous subsection is preserved here. Thus, in particular, the fact that the monopolist innovates only after resource exhaustion is not an argument against him. Considerations of efficiency dictate that this ought to be the sequence of events in this model. The inefficiency that the monopolist engenders is due both to the fact that his long run price, p^m, exceeds the unit cost of production *and* to the fact that during the resource-era the price does not, in general, rise at the rate of interest. This last has one exception, noted by Stiglitz (1976). This occurs if the market demand function is characterized by constant elasticity. In this case, if the elasticity of demand is greater than unity (i.e. $\eta < 1$), marginal revenue is proportional to price and, since the cost of extraction is by hypothesis negligible, the monopolist's extraction policy will follow the Hotelling Rule. But *Proposition 3* implies that even for this constant elasticity case the monopolist throughout sets a price in excess of the competitive price, since the long run price, p^m, that he *aims* at is higher. It follows therefore that the monopolist's depletion policy is excessively conservative. Transition to the new technology occurs later than is socially desirable: the backstop technology is held in abeyance for too long; see Fig. 1. Indeed, the fact that a monopolist throughout charges a price in excess of the competitive one can be shown to hold under more general circumstances; see Dasgupta & Stiglitz (1980). This is expressed in

Proposition 4. If $\eta(Q) < 1$ and $\eta'(Q) \geq 0$ (i.e. the elasticity of demand is a declining function of output), then the monopolist innovates later than is socially desirable and, in addition, the resource price that he sets prior to innovation is throughout in excess of the socially desirable level: see Fig. 1.

III.4. *Duopoly*

I now discuss a duopoly market, a case of considerable practical interest, and suggest that in the context of resource depletion the duopoly outcome under a large class of circumstances is even more conservative than pure monopoly;

or in other words, that in initial years the resource is sold at a price in excess of its monopoly price. Furthermore, we shall note that unlike the two polar market structures we have so far analysed, duopoly sustains a *gradual* transition from the exhaustible resource base to the inexhaustible one. Resource exhaustion does not precede technological innovation.

I assume now that the resource has been cartelized and that the backstop technology is protected by a patent held by a rival. For simplicity of analysis I take it that the patent is of sufficiently long duration to allow one to ignore the effect of the industrial organization after the patent expires. The aim now is to characterize an intertemporal game equilibrium in this setting. As regards the strategies chosen by the duopolists and the conjectures that they may reasonably hold, various possibilities suggest themselves. Here I shall restrict myself to the case where each of the players chooses the *quantity* he will supply over time and where each entertains Cournot conjectures regarding the strategy of the other. That is, each takes the quantities marketed by the other as given when choosing his own strategy.[1] It is now clear that so long as the resource owner is the sole supplier, his extraction policy must result in his marginal revenue rising at a percentage rate equal to r. Moreover, it is clear that subsequent to resource exhaustion the quantity marketed by the producer must yield him a marginal revenue equal to C; that is, the market price will be p^m after the resource has been exhausted. The interesting question arises whether the duopolists will ever supply simultaneously. Suppose they were to do so over some period. Then during this interval (a) the (net) marginal revenue earned by the resource cartel must rise at the rate of interest, r, and (b) the marginal revenue accruing to the rival must equal C. Thus let Q_t be the total quantity supplied to the market and let μ_t be the share of the market captured by the resource cartel at t. Then marginal revenue accuring to the resource cartel is

$$p(Q_t) + \mu_t Q_t p'(Q_t) = p(Q_t)[1 - \mu_t \eta(Q_t)], \tag{3a}$$

and the marginal revenue earned by the rival is

$$p(Q_t)[1 - (1 - \mu_t)\eta(Q_t)]. \tag{3b}$$

For simplicity of exposition suppose that the market demand curve is isoelastic. We must then suppose that $0 < \eta < 1$. Then, we have noted in (a) and (b) above that so long as $0 < \mu_t < 1$, an intertemporal Cournot equilibrium must be characterized by the two conditions

$$\dot{p}_t/p_t - \eta\dot{\mu}_t/(1 - \eta\mu_t) = r, \tag{4}$$

and

$$p_t[1 - (1 - \mu_t)\eta] = C. \tag{5}$$

(Notice that (5) implies that $C < p_t < C/(1-\eta)$.)

---

[1] Note in particular that I am implicitly supposing that the two players make their moves simultaneously. If we were to suppose that the resource cartel has the first move then the outcome that one would analyse is the leader-follower model of von Stakelberg. For an analysis of this see Dasgupta and Stiglitz (1980).

I now argue that the market share, μ_t, changes in such a way that both (4) and (5) hold over an interval of time. The idea is to eliminate both μ_t and $\dot{\mu}_t$. Thus differentiate equation (5) with respect to time. This yields

$$\dot{p}_t[1-\eta(1-\mu_t)]/p_t = -\eta\dot{\mu}_t. \tag{6}$$

Now use equation (6) in equation (4) to obtain the condition

$$\dot{p}_t/p_t = r[(1-\eta\mu_t)+C/p_t]/(1-\eta\mu_t). \tag{7}$$

Finally use equation (5) in equation (7) to obtain

$$\dot{p}_t/p_t = r(2-\eta-C/p_t)/(2-\eta). \tag{8}$$

Equation (8) characterizes the price trajectory under duopoly during the phase when the two share the market. The first point to notice about this equation is that $\dot{p}_t/p_t < r$; that is, market price rises at a percentage rate less than the rate of interest. Intuition also suggests that during this interval μ_t monotonically declines—that is, the share of the market controlled by the cartel falls. This is in fact the case.[1] But then one may conclude from equation (5) that even during this phase market price rises; i.e. $0 < \dot{p}_t/p_t < r$. In fact one can show that the price trajectory under duopoly is a continuous function of time. Thus in particular the date the duopoly price reaches the level $p^m (= C/(1-\eta))$, the resource is exhausted and the rival takes over the entire market. The question arises whether the duopolists share the market from the beginning, or whether there is an initial interval when the *cartel* is the sole supplier, setting its price *below* C, and thereby keeping the rival out of the market. Again, intuition suggests that if the resource stock is large such an initial phase will occur along a duopoly equilibrium, but that for small stocks the duopolists will share the market from the beginning. In fact it is possible to prove the following result.[2]

Proposition 5. If the initial resource stock, S_0, is large, a Cournot duopoly equilibrium is characterized by *three* phases. During the first phase (an interval $(0, T_1)$, where T_1 is endogenously determined), the cartel is the sole supplier. It extracts in such a manner that its marginal revenue (and since η is constant, the market price) grows at the rate r (equation (4) with $\mu_t = 1$). Furthermore the market price during this interval is less than the rival's production cost C. At T_1 the industry enters the second phase, with $p_{T_1} = C$, where during an interval (T_1, T_2), (with T_2 also determined endogenously) the duopolists share the market in such a manner that $\dot{\mu}_t < 0$, and equations (4) and (5) are both satisfied, and therefore, that $0 < \dot{p}_t/p_t < r$. At T_2 the resource is exhausted and $p_{T_2} = C/(1-\eta)$, and the industry enters the third and final phase where the

---

[1] See Dasgupta & Stiglitz (1976, 1980).
[2] See Dasgupta & Stiglitz (1976, 1980).

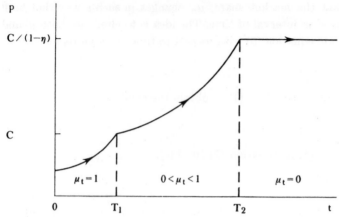

Fig. 2

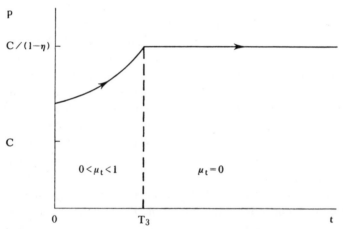

Fig. 3

rival controls the entire market and sets the price at $p^m = C/(1-\eta)$; see Fig. 2.[1] If the initial stock is "small" the first phase does not occur, and the duopolists share the market from the beginning; see Fig. 3. The price trajectory along an equilibrium path is a continuous function of time.

A comparison of Propositions 4 and 5 enable us to prove.[2]

Proposition 6. If the initial stock is large, the resource price under duopoly in initial years is in excess of its monopoly price. If the initial stock is small, the date of technological innovation is earlier, and the date of resource exhaustion is later under duopoly than they are under either perfect competition or monopoly.

---

[1] One can check by integrating equation (8) that the duration of the second phase, $T_2 - T_1 = 2/r \log (1-\eta)$.
[2] For a formal proof see Dasgupta & Stiglitz (1980).

It seems to me that *Proposition 6*, even though it refers only to the duopoly case, is disturbing. For it is intuitively clear that the biases the Proposition identifies are not restricted to duopoly but to a market in which there are only a *few* resource owners and a *few* manufacturers of the substitute product. The Proposition alerts us to the discomfiting fact that the polar market structures of perfect competition and monopoly do not offer bounds within which oligopolistic outcomes must necessarily lie. In particular, the oligopoly price of a natural exhaustible resource may well be in excess of its monopoly price. Moreover, resource substitutes may make their appearance earlier than is socially desirable, and exhaustible resources may be marketed for a longer period than under monopoly. We should note finally from Proposition 5 (and Fig. 2 and 3) that the intertemporal duopoly equilibrium we are discussing here is a *perfect equilibrium*, in the sense that at *each* date the remaining portion of the equilibrium outcome is an intertemporal duopoly equilibrium at that date. This is worth emphasizing since, if an intertemporal equilibrium is not perfect, it is not credible.[1]

Let $\pi(S_0)$ be the present-value of profits accuring to the owner of the backstop technology along the duopoly equilibrium and lets $R(S_0)$ denote the present-value of rents earned by the resource cartel. Then since we have been analysing a non-cooperative duopoly equilibrium, we know that for any value of S_0,

$$\pi^m(S_0) > \pi(S_0) + R(S_0), \tag{9}$$

(the sum of two duopolists profits must be lower than a pure monopolist's maximum profit).

We shall find (9) to be of importance in locating the *relative* incentives that agents have for undertaking resource-substitute inventions.

IV. Suppliers' Incentives for Developing Resource-Substitute Technologies

I come now to the issue I began with: the relationship between the pricing of exhaustible natural resources and the incentives for developing technologies that are designed to release an economy from the binding constraints such resources impose. In recent years "clean" fast-breeder reactors and controlled nuclear fusion have provided the best potential examples of the kind of construction we have been discussing in this paper. In what follows I present the very simplest of models to capture these issues. Complications can most cer-

---

[1] The concept of perfect (Nash) equilibrium, formalized by Selten (1975), is the game theoretic counterpart of the concept of *intertemporal consistency* of an individual agent's plan. This latter concept, discussed originally by Strotz (1955), has, most recently, been explored in a most engaging manner by Yaari (1978) in the context of the depletion of an appetite-arousing cake.

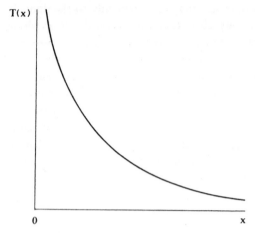

Fig. 4

tainly be added. But here I want merely to explore how the basic arguments would run. In this section I consider the case of a competitive patent race. In the next section I consider the socially optimal R & D programme for a resource importing country.

Suppose then that at $t=0$ the resource stock is S_0 and is controlled by a cartel. Furthermore, assume that there is no backstop technology available at this date. We now suppose that firms can undertake R & D expenditure to develop a backstop technology which will enable a substitute to be provided at a unit operating cost of $C>0$. What I want to capture here in a sharp manner is the *speed* with which this new technology will be developed under a specific industrial organization. Thus, I take it that if a typical firm invests an amount x at $t=0$, the date at which it will have solved all the R & D problems is guaranteed to be $T(x)$, where $T'(x)<0$. That is, the greater the R & D expenditure a firm chooses to incur, the earlier it will have developed this new technology. In particular, I am supposing here that there is no uncertainty in the R & D process. To simplify the exposition I also suppose that the R & D technology is characterized by the conditions $T''(x)>0$, $T(0)=\infty$ and $T(\infty)=0$ see Fig. 4.[1] Now, in order that firms have an incentive to develop the backstop technology they must be able to exploit some monopoly power over their invention. To capture this sharply I take it that the first firm to develop the backstop technology is awarded the patent and that the patent provides per-

---

[1] The R & D process envisaged here is, of course, very stylized; for it supposes that a new technology, known in advance, becomes viable in an abrupt manner once a set of research problems is solved. But it ought also to be clear why, for the questions being raised in this paper these simplifications are desirable ones to make. In his thoughtful review of Dasgupta & Heal (1979), in the *Economic Journal* (1980), professor Murray Kemp has raised a perceptive point, to the effect that in the world as we know it, R & D processes are not quite like this.

fect protection. To simplify again, I take it that the patent is of sufficiently long duration to allow us to ignore the period after the patent expires. In short, I suppose that the winner takes all.

Now suppose for the moment that for one reason or the other the resource owner is prohibited from entering the patent race. This may be because there are strong anti-trust laws which the resource owner fears. Thus, if there is a *single* winner of the patent race (i.e. there is no tie), the postinvention era will be characterized by a *duopoly* industrial structure, whose outcome we have already analysed under certain assumptions. In particular, if a firm wins the patent race at date T, and the resource stock remaining at T is $\bar{S}$ (with $S_0 \geqslant \bar{S} \geqslant 0$), the present-value of *profits* that the patent winner earns from that date can be computed from the analysis of the previous section; see *Proposition 5*. As in Section III, let $\pi(\bar{S})$ denote this present-value, computed from the vantage point of date T. It is an easy matter to check from the analysis leading to *Proposition 5* that $d\pi(\bar{S})/d\bar{S} < 0$. Which brings me back to the point I emphasized in the Introduction that, other things remaining the same, the reward to be had by the firm that develops a backstop technology depends on the size of the stock remaining at the date R & D is completed. Furthermore— and this is especially important—for the model under consideration, the *larger* the size of the remaining stock, the *smaller* the reward. It follows that the size of the remaining reserves at each date is a strategic variable from the *cartel's* point of view. We now explore this more formally.

As we are supposing that the research sector is competitive it makes good sense to assume that the resource cartel makes the first move and announces its pricing—extraction strategy in full knowledge of the fact that any credible announcement on its part will elicit a response from the research sector by way of R & D expenditure, and hence by way of an invention date. To see how this response may be analysed, suppose the cartel announces a pricing/extraction strategy for the pre-invention era which implies that at each date $T (\geqslant 0)$ the remaining stock will be $\bar{S}(T)$. For simplicity of exposition I suppose that all firms—barring the cartel, which by hypothesis is not allowed to compete— face identical research technologies given by the function $T(x)$. Thus, if a firm invests x in R & D, and if this is in excess of what any other firm has chosen to spend, then it wins the patent at date $T(x)$, and the present-value of its *net* profits may then be expressed by the formula

$$\pi(\bar{S}(T(x)))e^{-rT(x)} - x. \tag{10}$$

Now clearly an equilibrium *in the research sector* will not sustain any losers, since we are supposing that the winner takes all. Moreover, an equilibrium cannot sustain a tie for the patent, because in case of a tie the joint winners will *share* the spoils and in this case any one of the joint winners—making Cournot hypotheses regarding the others—can do better by spending only a little more on R & D, thus guaranteeing the patent exclusively for itself. Thus

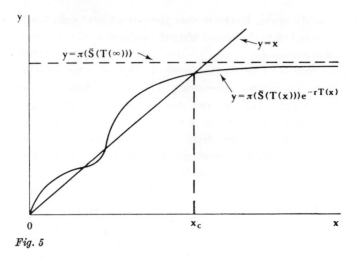

$y = \pi(\bar{S}(T(\infty)))$

$y = x$

$y = \pi(\bar{S}(T(x)))e^{-rT(x)}$

x_c

0 x

Fig. 5

an equilibrium in the research sector sustains at most one firm. Since there is always the *threat* of entry into the research sector one concludes that, given the function $\bar{S}(T)$, equilibrium R & D expenditure in the research sector, which I denote by x_c is given by the *largest* solution of the *zero-profit condition*

$$\pi(\bar{S}(T(x)))e^{-rT(x)} = x, \tag{11}$$

(see Fig. 5).

Equation (11) yields the *reaction function* of the research sector. I turn now to pricing/extraction strategy of the cartel faced with such a reaction function. The analysis is vastly simplified if we suppose, as in Section III.4, that the market demand curve is iso-elastic and that $\eta < 1$. In this case it is plain that *until* the *invention* is made the cartel will so wish to extract as to ensure that market price rises at the rate of interest r. What remains to be determined is the choice of the *initial* price by the cartel. Now, suppose the cartel sets the initial price at p_0. Then, at all t prior to invention, $p_t = p_0 e^{rt}$. It follows from the market demand function $p(Q_t)$ that $Q_t = p^{-1}(p_0 e^{rt}) \equiv f(p_0 e^{rt})$, say. It follows that

$$\bar{S}(T) = S_0 - \int_0^T f(p_0 e^{rt}) \, dt. \tag{12}$$

Substituting (12) in (11) we obtain

$$\pi\left(S_0 - \int_0^{T(x)} f(p_0 e^{rt}) \, dt\right) e^{-rT(x)} = x. \tag{13}$$

Thus, for any choice of p_0 by the cartel, the research sector responds by engaging in R & D expenditure of the amount x_c, which is the largest solution of the zero-profit condition (13). Thus in particular, $x_c = x_c(p_0)$, and it is p_0 which is chosen by the cartel. We turn to this choice.

Suppose T is the date of invention, and suppose $\bar{S}(T)$ is the cartel's reserve at that date. As in Section III.4, let me denote by $R(\bar{S}(T))$ the present-value of *rents* that the cartel earns from T onwards along the Cournot duopoly equilibrium, where $R(\bar{S}(T))$ is computed from the vantage point of time T. But we have already concluded that for any choice of p_0, equilibrium R & D expenditure is $x_c(p_0)$, and therefore the date of invention is $T(x_c(p_0))$. Thus, the present-value of rents accruing to the cartel, as a function of its choice of initial price, p_0, is

$$V(p_0, T) \equiv \int_0^{T(x_c(p_0))} (p_0 e^{rt}) f(p_0 e^{rt}) e^{-rt} \, dt + R\left(S_0 - \int_0^{T(x_c(p_0))} f(p_0 e^{rt}) \, dt\right) e^{-rT(x_c(p_0))},$$

(14)

and the cartel chooses p_0 so as to maximize this. Let this optimum initial price be written as p_0^*. It follows that p_0^* must satisfy the first-order condition

$$\partial V/\partial p_0 + (\partial V/\partial T) \, T'(x_c) \, (dx_c/dp_0) = 0. \tag{15}$$

This concludes our description of how resource extraction and R & D expenditure are related along an intertemporal oligopoly equilibrium. It will, however, be noticed from the foregoing analysis that equilibrium p_0, i.e. p_0^*— obtained from equation (15)—is *higher* than the equilibrium initial price that would have prevailed *were* the cartel to suppose that its pricing policy does not affect R & D expenditure by rival firms. This merely re-affirms my earlier contention that, contrary to what is often thought, the setting of a high price for a resource acts as a dampener on the incentives for R & D expenditure by rival firms. But if the cartel sets its initial price too high its revenue in initial years is too low. It is a delicate balancing of these two considerations that determines the cartel's optimum price policy. Equation (14) captures these considerations for the model at hand. Notice also from the way the oligopoly equilibrium has been constructed that the equilibrium is *perfect* in the sense of Selten (1975).

I want next to draw attention to the fact that if S_0 is "large" *and* if R & D technological possibilities, as represented by the function $T(x)$, are "favourable", then $x_c(p_0^*)$ is "large" and hence $T(x_c(p_0^*))$ is "small", so that $\bar{S}(T(x_c(p_0^*))) \equiv S_0 - \int_0^{T(x_c(p_0^*))} f(p_0^* e^{rt}) \, dt$, is not too small relative to S_0. This means, as is implied by Fig. 2 and *Proposition 5*, that there is an initial interval *subsequent* to the date of invention during which the cartel is the sole supplier and sets a price below C. That is, the patent holder will not have innovated until some time after he has captured the patent and is capable of innovating. Such a "sleeping patent" is clearly socially wasteful, for the firm could spend less on R & D without affecting the date of innovation. But the patent winner chooses not to delay winning the patent simply because if he were to attempt to do so some other firm would enter the race and defeat him.[1]

The forces of R & D competition dissipate profits, and this is reflected in equation (13). It will be recalled, however, that I have supposed that the cartel is unable to participate in the patent race. I shall now argue that it is the resource cartel that has the greatest incentive to acquire the patent if it is allowed to do so. The intuitive reason is not difficult to see. If the resource cartel were also to own the patent it would be a pure monopolist. As before, let $\pi^m(\bar{S})$ denote the present-value of maximum profits earned by such a monopolist if $\bar{S}$ is his resource stock at the date he acquires the patent (see *Proposition 3*). From (9) we may conclude

$$\pi^m(\bar{S}) > \pi(\bar{S}) + R(\bar{S}). \tag{16}$$

But from (13) we know that because of competition the winning firm earns no profit. Thus suppose the cartel were to finance the winning firm's R & D expenditure $x_c(p_0^*)$ with the proviso that the patent right will be given over to it. Now if we write $\bar{S}(p_0^*) \equiv S_0 - \int_0^{T(x_c(p_0^*))} f(p_0^* e^{rt})dt$, then (16) implies that

$$\pi^m(\bar{S}(p_0^*)) e^{-rT(x_c(p_0^*))} - x_c(p_0^*) > \pi(\bar{S}(p_0^*)) e^{-rT(x_c(p_0^*))} - x_c(p_0^*) + R(\bar{S}(p_0^*)) e^{-rT(x_c(p_0^*))}, \tag{17}$$

which, on using (11) implies that

$$\pi^m(\bar{S}(p_0^*)) e^{-rT(x_c(p_0^*))} - x_c(p_0^*) > R(\bar{S}(p_0^*)) e^{-rT(x_c(p_0^*))}. \tag{18}$$

We conclude that it pays the cartel to win the patent.[2]

Inequality (18) shows that if the cartel is legally allowed to hold the patent it will acquire it. But the foregoing analysis does not yield the *intertemporal equilibrium were* it allowed to acquire it. I turn, therefore, to locating an equilibrium in the case where the cartel is *not* barred from competing for the patent. This is an easy matter. For, since we know that the cartel will in fact acquire the patent we know that subsequent to the invention the entire market will be controlled by it. Proposition 3 is relevant now for the post-invention market. Thus we must suppose that the cartel chooses both the *initial* price, p_0, *and* R & D expenditure x. However, since the research sector is, by hypothesis competitive, the cartel must choose p_0 and x subject to the constraint $x \geqslant x_c(p_0)$, where $x_c(p_0)$ is the largest solution of equation (13). Observing this constraint on R & D expenditure enables it to acquire the patent. We conclude therefore that if the cartel can compete for the patent, then on our assumption

---

[1] For a rigorous presentation, see Dasgupta, Gilbert & Stiglitz (1980).
[2] For a discussion of this issue in a different context, see Dasgupta & Stiglitz (1980a).

that the cartel makes the first move in our oligopoly game with no pre-play communication, the intertemporal equilibrium is given by the solution of the cartel's maximization problem, which is:

to choose p_0 and x so as to

$$\text{maximize} \int_0^{T(x)} f(p_0 e^{rt})\,(p_0 e^{rt})\,e^{-rt}\,dt + \pi^m \left(S_0 - \int_0^{T(x)} f(p_0 e^{rt})\,dt \right) e^{-rT(x)} - x \qquad (19)$$

subject to the constaint $x \geqslant x_c(p_0)$, where $x_c(p_0)$ is the largest solution of the research sector's zero-profit condition (13).

Notice first that the solution of problem (19) describes a *perfect* equilibrium in the sense of Selten (1975): at each date the remaining portion of the solution of (19) is an intertemporal equilibrium at that date. Notice also that if S_0 is large and if the cartel's optimum x in (19) is large the cartel also will find it in its interest to hold a sleeping patent for a period after it has acquired the patent. The reason is the same as before: holding a sleeping patent is the "price" the cartel has to pay to *obtain* the patent. Otherwise, someone else will acquire it.

We may now summarize the main conclusions in the form of

Proposition 6. A resource cartel has greater incentives than anyone else to acquire the patent to rival technology. If there is competition in R & D the industrial organization may well sustain a sleeping patent—that is, the development of an alternative technology well before it makes its appearance in the market. This may happen whether or not the cartel gains control of the technological invention.

The phenomenon of sleeping patents in this model is socially wasteful. In the world as we know it the phenomenon may well occur even under the best circumstances simply because one cannot predict the date of success of a research programme. But in this paper I have eschewed such uncertainty. The driving force behind excessive R & D investment—in the sense of preemptive patenting—is *competition* in the patent race. What is worth bearing in mind in this case is that the winning form knows in advance that it will be unprofitable for it to innovate for a while after its R & D programme is successfully completed: for a period the resource cartel will undercut the patent winner's new technology.

V. Optimum R & D Policy for a Resource Importing Economy

V.1. *The Framework*

In the previous section I was concerned with *suppliers'* incentives for developing alternative technologies in the face of a resource cartel. I was simultaneously concerned with the cartel's pricing strategy which, in an intertemporal oligopoly equilibrium, is related to such incentives. I now wish to sketch the kinds of considerations that are involved in assessing resource *users'* incentives for developing alternative technologies in the face of a resource cartel; for this I shall consider arguments that are involved in devising an *optimal* R & D programme for a nation that relies exclusively on a foreign cartel for the supply of a resource.[1] For simplicity I take it that the cartel also supplies exclusively to this importing country. I shall be supposing that the government of the importing country chooses the domestic R & D programme which, in our model, means the choice of R & D expenditure x. The choice of x results, with complete certainty, in the completion date $T(x)$. For simplicity I assume that the government will not be involved in actual production when the backstop technology is developed. Rather, it will be supposed that the government will make publicly available information about the new technology. Thus, while I suppose that the government chooses the level of R & D expenditure, I also suppose that on the completion of the R & D programme the new technology will be competitively owned.

5.2. *The Argument*

It is simplest, in analysing the above problem, to consider first the simpler one in which the date of *invention* of the backstop technology is *exogenously* given. Suppose this date to be $T(\geqslant 0)$. We may interpret this as being a situation in which the government of the importing country announces that it will make publicly available information bearing on the future backstop technology at date T. It must obviously be supposed that such an announcement is backed by some form of evidence, so that the announcement is credible.

This mixed market structure, in which the resource is controlled by a cartel and in which from some future date T there is a competitively owned backstop technology, is a variation of a model studied by Hoel (1978).[2] Since the rival technology will not be available until T the cartel can set its price any way it likes until then. However, from T onwards it is constrained to set its price *below or equal* to C, the unit cost of production associated with the backstop. Thus, if we let $Q = f(p)$ denote the *inverse* of the importing nation's demand function, the cartel is forced to observe the constraint $Q_t \geqslant f(C)$ for $t \geqslant T$.[3] We suppose that the cartel's goal is to maximize the present-value of its rents. Thus suppose first that the cartel's reserves at T amount to $\bar{S}$ and now define

---

[1] For a more complete discussion, see Dasgupta, Gilbert & Stiglitz (1981).

[2] Hoel (1978), in his excellent study, considered the special case $T = 0$.

[3] In other words, we are making the innocuous assumption that the cartel can serve the entire market of size $f(C)$ if it sets its price at C. We shall note the occurrence of the *limit pricing* phenomenon in what follows.

$$\left.\begin{array}{l} \tilde{R}(\bar{S}) = \max \displaystyle\int_{T}^{\infty} p(Q_t)\, Q_t\, e^{-r(t-T)}\, dt \\[4mm] \text{subject to } \displaystyle\int_{T}^{\infty} Q_t\, dt = \bar{S}, \quad Q_t \geqslant 0 \quad \text{and} \quad Q_t \geqslant f(C). \end{array}\right\} \tag{20}$$

$\tilde{R}(\bar{S})$ is therefore the maximum present-value of rents that the cartel can earn from T onwards if $\bar{S}$ is the size of its reserves at T. Problem (20) has been analysed by Hoel (1978). Now it will be noticed that we have defined $\tilde{R}(\bar{S})$ from the vantage point of T. It follows then that the cartel's *full* problem is:

$$\left.\begin{array}{l} \text{to choose } Q_t(T \geqslant t \geqslant 0) \quad \text{and} \quad \bar{S}(S_0 \geqslant \bar{S} \geqslant 0) \quad \text{so as to:} \\[4mm] \text{maximize } \displaystyle\int_{0}^{T} p(Q_t)\, Q_t\, e^{-rt}\, dt + \tilde{R}(\bar{S})\, e^{-rT} \\[4mm] \text{subject to } \displaystyle\int_{0}^{T} Q_t\, dt = S_0 - \bar{S}, \quad Q_t \geqslant 0. \end{array}\right\} \tag{21}$$

The cartel's problem, as formalized in (21), may be analysed by routine techniques. For simplicity I consider the case where the importing country's demand function is iso-elastic, with the elasticity of demand in excess of unity (i.e. $0 < \eta < 1$). In this case the solution of (21) can be expressed in the form of:

Proposition 7. There exists a function $T^*(S_0)$ (with $dT^*(S_0)/dS_0 \geqslant 0$) such that if $0 \leqslant T \leqslant T^*(S_0)$, there exist two instants, T_1 and T_2 (with $T_1 = T^*(S_0)$ and $T_2 \geqslant T_1$) such that the cartel is the sole supplier during $(0, T_2)$. During $(0, T_1)$ the cartel so controls its supply that the market price rises at a percentage rate equal to r. It chooses its initial price p_0 in such a manner that $p_0 e^{rT_1} = p_{T_1} = C$. During (T_1, T_2) the cartel markets its remaining stock at the price C. At T_2 the stock is exhausted and the competitively owned backstop technology makes its appearance and for $t \geqslant T_2$ the product is sold at the price C. Furthermore, market price is a continuous function of time; see Fig. 6.[1] If $T > T^*(S_0)$, there exists an instant T_3 (with $T_3 \geqslant T$) such that the cartel is the sole supplier during $(0, T_3)$. During $(0, T)$ the cartel so controls its supply that the market price rises at a percentage rate equal to r. The initial price p_0 is so chosen that $p_0 e^{rT} > C$. At the date of invention T, there is a discontinuous fall in the price, and the cartel proceeds to market its remaining stock at the price C. At T_3 the stock is exhausted and the competitively owned backstop technology makes its appearance and for $t \geqslant T_3$ the product is sold at the price C; see Fig. 7.

---

[1] This part of the Proposition is proved in Hoel (1978) for the special case $T = 0$.

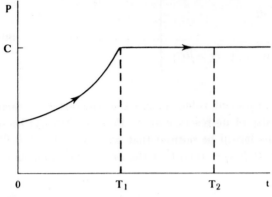

Fig. 6

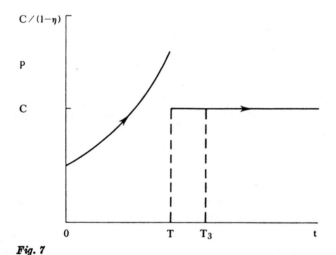

Fig. 7

Let $U(Q) = \int_0^Q p(Q')dQ'$, be the social surplus enjoyed by the importing country when Q is total consumption. I shall suppose that the flow of social welfare for the importing country is measured by the flow of net social surplus, $U(Q_t) - p(Q_t)Q_t$, and that the government measures social welfare by its present discounted value, and I take it also, for simplicity, that the importing nation's social rate of discount equals the competitive rate of interest r. If T is announced by the importing country to be the date at which its R & D programme will be completed, then the cartel, in our model, chooses its export policy by solving (21). Thus let Q_t^* denote the cartel's rent-maximizing extraction policy and let $\bar{S}^*(T)$ be the stock of reserves it plans to hold at T along the policy. Since import demand is by hypothesis iso-elastic, we know that $Q_t^* = f(p_0 e^{rt})$ during $0 \leqslant t \leqslant \max \{T, T^*(S_0)\}$, (see Figs. 6 and 7), and that $Q_t^* = \bar{Q}$ thereafter until reserves run out, where $\bar{Q}$ is the solution if $p(Q) = C$. In particular, we know from *Proposition 7* that $\bar{S}^*(T) > 0$ if T is "small", but that $\bar{S}^*(T) = 0$ if T is "large". Let $Z(T, S_0)$ denote the present value of net social surplus enjoyed by the importing nation from T. Then we observe from *Proposition 7* that

$$Z(T, S_0) = \bar{Z} = \int_T^\infty [U(\bar{Q}) - C\bar{Q}] e^{-r(t-T)} dt = [U(\bar{Q}) - C\bar{Q}]/r, \quad \text{if } T > T^*(S_0),$$

and $\qquad (22)$

$$Z(T, S_0) = \int_T^{T^*(S_0)} [U(f(p_0 e^{rt})) - p_0 e^{rt} f(p_0 e^{rt})] e^{-r(t-T)} dt$$

$$+ \int_{T^*(S_0)}^\infty [U(\bar{Q}) - C\bar{Q}] e^{-r(t-T)} dt, \quad \text{if } T < T^*(S_0). \qquad (23)$$

It follows that we may represent the importing nation's social welfare by the expression:

$$W(T, S_0) = \int_0^T [U(f(p_0 e^{rt})) - p_0 e^{rt} f(p_0 e^{rt})] e^{-rt} dt + Z(T, S_0) e^{-rt},$$

and it will be noted that $\partial W/\partial S_0 > 0$ and $\partial W/\partial T \leq 0$. But $T = T(X)$. Assuming that R & D investment can be met from general taxation social welfare, net of such investment, can be expressed as:

$$W(T(x), S_0) - x = \int_0^{T(x)} [U(f(p_0 e^{rt})) - p_0 e^{rt} f(p_0 e^{rt})] e^{-rt} dt$$

$$+ Z(T(x), S_0) e^{-rt(x)} - x, \qquad (24)$$

where $p_0 = p_0(T(x))$, as in *Proposition 7*. It follows that the importing government's planning problem is to:

$$\left. \begin{array}{l} \text{choose} \quad x \geq 0 \quad \text{so as to} \\ \text{maximize } \{W(T(x), S_0) - x\}. \end{array} \right\} \qquad (25)$$

Now, we know from *Proposition 7* that $dp_0/dT = 0$ if $T < T^*(S_0)$. It follows immediately that the government will *not* choose $T < T^*(S_0)$: there is no gain in having the invention before $T^*(S_0)$; (see also (23)). Let $\bar{x}$ be the solution of $T(x) = T^*(S_0)$. Then we conclude that (25) may as well be written as:

$$\left. \begin{array}{l} \text{choose} \quad x \leq \bar{x} \quad \text{so as to} \\ \text{maximize } \int_0^{T(x)} [U(f(p_0 e^{rt})) - p_0 e^{rt}]) e^{-rt} dt \\ \quad + e^{-rT(x)} [U(\bar{Q}) - C\bar{Q}]/r. \end{array} \right\} \qquad (26)$$

It follows from (26) that Fig. 6 will not be observed; but only Fig. 7.

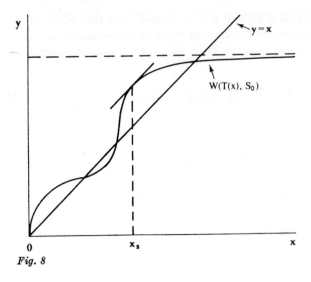

Fig. 8

It is important to emphasize that despite the strong regularity conditions that we have so far imposed on the functions $p(Q)$ and $T(x)$, it is entirely possible that W in (25) is *not* a concave function of x, and, in particular, that $\{W(x) - x\}$ possesses a number of local maxima and minima; see Fig. 8. In such a case it would be treacherous to rely on *marginal* social cost-benefit analysis to locate the optimal level of R & D. Let x_s be the solution of (26).

Now clearly $x_s = x_s(S_0)$: optimal R & D investment depends on the cartel's resource base. In particular, for a wide class of R & D technologies, $T(x)$, it is possible to show that $x_s(S_0) = 0$ if S_0 is "large" and $x_s(S_0) > 0$ for S_0 "small". What this says is precisely what one would expect: if the cartel's stock is large there is no point in embarking on an R & D programme now. This does not mean that the government ought *never* to incur R & D expenditure; merely that it ought to wait until the cartel has depleted its stock somewhat. But perhaps the most interesting implication of the analysis is that for a wide class of parametrizations the maximization of (26) yields a level of optimal R & D expenditure, $x_{s'}$ for which $\bar{S}^*(T(x_s)) > 0$. In this case the importing country's government knows, at the time of setting the pace of research and development that the technological breakthrough will occur at a date $T(x_s)$ at which the cartel will still have reserves left, and knows therefore, that the cartel will continue to supply the market for a period after this date. In short $\bar{S}^*(T(x_s)) > 0$ means that technological *innovation* will take place only some time after the optimum date of invention $T(x)_s$. In order to make *credible* the announcement that it will have made the technological breakthrough by a certain date it must actually undertake the research programme with the appropriate intensity. Thus, it may well be in the country's interest to embark on a *crash programme*, one which involves its making the technological breakthrough some time before it plans to use it.

Finally, a comparison of equations (11) and (26) suggests that the competitive rate of R & D expenditure, x_c, may be greater or less than the socially optimal level x_s. When $x_c > x_s$ it is simply that free entry forces the innovating firm to make the invention even earlier than is socially desirable.

I summarize these considerations in the form of

Proposition 8. Even in the absence of uncertainty, optimal R & D for a resource importing country may involve a crash programme, which is to say that the programme may involve developing a backstop technology some time before it will actually be used. Furthermore, marginal social costbenefit analysis is unreliable as a procedure for locating the optimal R & D programme. In general global social cost-benefit analysis is required.

VI. Commentary

In this paper I have, in the context of a simple construction, made explicit the strategic roles that are played both by the pricing of natural resources by cartels and by the choice of R & D programmes by rivals desiring to develop substitute technologies. My main aim has been to draw out the intertemporal structure that such oligopoly games must have. The model used to examine these questions is merely an example; nothing more. But it is rich enough to lay bare what seems to me to be the most important considerations.

The fact that R & D processes are shot through with uncertainty is something that can be accommodated within the framework studied here; so too can "political goals", and in particular that an importing nation may fear embargoes. But this latter—providing, as it does an *insurance* value to R & D expenditure—merely *reinforces* the case for crash programmes, the possible desirability of which is highlighted by *Proposition 8* even when there are *no* threats of embargo.

In this paper my focus of attention has been directed at the incentives that exist for the development of substitute products, and as ought to be transparent, the examples I have looked at have been motivated very much by concern about the *energy* sector. But in order to concentrate attention on the R & D sector I have consciously simplified the market structure for the *exhaustible* resource. Thus, for example, the simulation exercises of Cremer and Weitzman (1976) and Pindyck (1978), and the theoretical explorations of Gilbert (1978) and, more recently, Ulph (1980) have, rightly, noted that exhaustible resources (e.g. crude oil) are often only *partially* cartelized and that the resource cartel, in choosing its pricing strategy, must take into account the response of the non-cartel resource owners—the "fringe".[1] At the conceptual

---

[1] Pindyck (1978) estimates that the cartelization of the oil market results in a 55 % increase in the present value of rents accruing to OPEC members—implying thereby that the incentives for the cartel to remain cohesive rare rather strong.

level this additional feature poses no serious difficulty. But, as Ulph (1980) has noted, *characterization* results are harder to come by, since an additional issue that is faced by the cartel is *when* to enter the market. At the practical stage of modelling these features further complications arise out of the fact that extraction costs vary across owners and that, in particular, capacity constraints simply cannot be ignored. That is to say, theoretical explorations to date have, for the most part, modelled the technology of extraction in a simplistic manner. Investment costs have been ignored; an exception is Jacobson & Sweeney (1979).

Among the remaining simplifications made in the basic model I have discussed here, one that merits particular attention is the fact that the coalitional structure of the resource cartel has been constructed only casually. The fact is, of course, that the goals of cartel members typically vary, and in this case the cartel's strategy will emerge only as a result of bargaining among themselves with the proviso that commitments among them are binding. Hynylicza & Pindyck (1976) were the first to broach this issue by an appeal to the Nash bargaining solution among cartel members where the goals among the members were assumed to differ by way of the discount rate they used for computing the present-value of rents. In a recent study Bhaduri & Mukherji (1979) recognize that cartel members often differ by way of their intentions with regard to how they propose to *use* their rent earnings. In the absence of competitive capital markets a difference in their intentions will plainly matter. Bhaduri & Mukherji also appeal to the Nash bargaining solution for arriving at the cartel's cooperative strategy. It should, however, be plain that introducing these considerations in developing the "cartel's goals" would not affect the conclusions that I have highlighted in this paper. It is for this reason that I have simplified and supposed that the cartel's goal is the maximization of the present-value of its profits—the present-value being computed by the use of a constant rate of interest r.

References

Bhaduri, A. & Mukherji, A.: OPEC and the oil price rise: A game theoretic approach. Centre for Economic Studies and Planning, Jawaharlal Nehru University, 1979 (mimeo).

Cremer, J. & Weitzman, M.: OPEC and. the monopoly price of world oil. *European Economic Review 8*, 155–164, 1976.

Dasgupta, P.: *The social management of environmental resources* (forthcoming). Basil Blackwell, 1982.

Dasgupta, P., Gilbert, R. & Stiglitz, J.: Invention and innovation under alter-

native market structures: The case of natural resources. Princeton University, 1980 (mimeo).

Dasgupta, P., Gilbert, R. & Stiglitz, J.: The social reward from inventive activity: The case of natural resources. London School of Economics, 1981 (mimeo).

Dasgupta, P. & Heal, G. M.: *Economic theory and exhaustible resources*, James Nisbet and Cambridge University Press, Welwyn and Cambridge, 1979.

Dasgupta, P. & Stiglitz, J.: Uncertainty and rates of extraction under alternative institutional arrangements. IMSSS Technical Report No. 179, Stanford University, 1976.

Dasgupta, P. & Stiglitz, J.: Market structure and resource depletion: A contribution to the theory of intertemporal monopolistic competition. *Journal of Economic Theory*, 1980 (forthcoming).

Dasgupta, P. & Stiglitz, J.: Uncertainty, market structure and the speed of R & D. *Bell Journal of Economics*, pp. 1–28, Spring 1980 a.

Dasgupta, P. & Stiglitz, J.: Resource depletion under technological uncertainty. *Econometrica 49* (1), 85–104, 1981.

Gilbert, R.: Dominant firm pricing in a market for an exhaustible resource. *Bell Journal of Economics*, Autumn 1978.

Hoel, M.: Resource Extraction, Substitute Production and Monopoly, *Journal of Economic Theory 19* (1), 28–37, 1978.

Hynlicza, E. & Pindyck, R.: Pricing policies for a two-part exhaustible resource cartel: The case of OPEC. *European Economic Review 8*, 139–154, 1976.

Jacobson, A. & Sweeney, J.: Optimal investment and extraction for a depletable resource. Stanford University, 1980 (mimeo).

Khalatbari, F.: Market imperfections and optimal rate of depletion of an exhaustible resource. *Economica 44*, 409–414, 1977.

Koopmans, T. C.: The transition from exhaustible to renewable or inexhaustible resources. In *Economic Growth and Resources*, (ed. C. Bliss and M. Boserup), vol. 3. Macmillan Press, London, 1980.

Pindyck, R.: Gains to producers from the cartelization of exhaustible resources. *Review of Economics and Statistics 60*, 238–251, 1978.

Selten, R.: Re-examination of the perfectness concept for equilibrium points in extensive games. *International Journal of Game Theory 4*, 25–55, 1975.

Stiglitz, J.: 'Monopoly and the rate of extraction of exhaustible resources. *American Economic Review 66* (4), 655–661, 1976.

Strotz, R.: Myopia and inconsistency in dynamic utility maximization. *Review of Economic Studies 23*, 165–180, 1955.

Thomas, Brinley: Towards an energy interpretation of the industrial revolution. *Atlantic Economic Journal 8* (1), 1–15, 1980.

Ulph, A. M.: Modelling partially cartelized markets for exhaustible resources. Department of Economics, University of Southampton, 1980 (mimeo).

Yaari, M.: Consistent utilization of an exhaustible resource; or, how to eat an appetite-arousing cake. Research Memorandum No. 26, Center for Research in Mathematical Economics and Game Theory, Hebrew University, Jerusalem, 1977.

MARKET STRUCTURE AND RESOURCE EXTRACTION UNDER UNCERTAINTY*

Joseph E. Stiglitz

Princeton University, Princeton, New Jersey, USA

Partha Dasgupta

London School of Economics, London, England

Abstract

This paper compares the rate of extraction of a natural resource under alternative market structures when there is uncertainty about the date of discovery of a substitute (or about the date of discovery of a new deposit). The analysis shows that imperfectly competitive market structures are excessively conservationist: for any value of the initial stock of the natural resource, the price is higher than in competitive equilibrium (and therefore higher than the socially optimal level). But markets with limited competition may have a higher price than markets with pure monopoly (the same monopolist controlling the natural resource and its substitute). In particular, we find that the highest prices are associated with (Nash quantity setting) duopolists; the next highest prices occur in markets in which the monopolist controls the resource but the substitute is competitively produced; the pure monopolist sets his price lower than this, while the market in which a monopolist controls the substitute but the resource is competitvely owned is still lower.

One of the strategies commonly proposed for responding to the OPEC cartel is for the consuming nations to develop substitutes which they could then ensure would be competitively supplied. Our analysis suggests that the response of the resource monopolist to this change in market structure would be to raise his prices, not to lower them.

I. Introduction

In this paper we compare the rate of extraction of a natural resource when there is uncertainty about the date of discovery of a substitute[1] under alternative market structures.

---

* This paper is a revised version of results originally reported in Dasgupta & Stiglitz, "Uncertainty and rates of extraction under alternative institutional arrangements", IMSSS Technical Report No. 179, Stanford, 1976. An earlier draft of this paper was completed while Stiglitz was Oskar Morgenstern Distinguished Research Fellow at Mathematica and visiting professor at the Institute for Advanced Study. Financial support from the National Science Foundation is gratefully acknowledged.
[1] The effect of uncertainty about the date of discovery of a new deposit may be analyzed in a very similar manner.

In an earlier paper, we compare the rate of extraction (and the date of innovation of the substitute) *after* the discovery occurs; see Stiglitz & Dasgupta (forthcoming). In this paper, we are concerned with the rate of extraction *prior* to the discovery of the substitute.

In our earlier paper, we were able to show how, under each of a variety of market structures, the price at the date of invention could be determined; for instance, the price in monopoly, $p^m(S)$, for every value of the stock, S, of the resource, was shown always to be greater than the price in competition (or the socially optimal price), for the corresponding value of the stock, $p^s(S)$. The fact that once the new substitute has been developed the price will be higher at each value of S naturally leads to the conjecture that *prior* to the invention, since the invention reduces the value of the natural resource, the resource will be used up less quickly, i.e. *prior* to the invention, the price will be higher, at each value of the stock.

This conjecture turns out to be correct for the particular comparison we have just made—between monopoly and pure competition. Thus, we are able to extend the result of our earlier analysis showing that monopolies have a bias to excessive conservationism to the period prior to the invention having occurred.

The conjecture, however, is not true generally: one cannot infer from the comparison of post-invention prices what pre-invention extraction rates will be.

In our earlier analysis, we showed that in some models of *limited competition*[1] the equilibrium did not lie between the polar cases of pure competition and pure monopoly. For these same models of limited competition, we show again that the rate of extraction need not be between those for the two "polar" cases of pure monopoly and pure competition. Moreover, even if the price immediately after the invention is less than in the pure monopoly, *prior* to the invention the market may be *more* conservationist *even than pure monopoly*.

The reason for these unexpected results is simple. A monopolist owning a natural resource in deciding his optimal extraction policy is concerned with the marginal revenue of an extra unit of the resource after the invention has been discovered. The relationship between marginal revenues and prices may differ markedly in different market structures. In particular, although the post invention price may be lower when the monopoly resource owner faces a competitive substitute than when the monopolist also controls the substitute, his marginal revenue may be higher, and it is this which leads him in the preinvention period to charge higher prices for the natural resource.

The actions which the owners of resources can take prior to the discovery

---

[1] We use the term limited competition to describe market structures in which there is more than one firm but at least one firm can affect market price significantly.

of the substitute have important implications for the timing of *innovation*[1] and, where the discovery date of the invention itself is affected, to some extent, by expenditures on R & D, on the level of those expenditures and the development of substitutes; in particular, our analysis suggests that certain forms of *limited competition* may be associated with particularly long lags in innovation. These questions are discussed in greater detail in Dasgupta, Gilbert & Stiglitz (1980*a*, 1980*b*), Stiglitz, Gilbert & Dasgupta (1978) and Dasgupta & Stiglitz (1980).

We examine in this paper five market structures:

(*a*) pure competition, or, equivalently, the socially optimal allocation of resources;

(*b*) pure monopoly, where the same firm owns both the resource and its substitute;

(*c*) competitive resource market, monopoly control of substitute; this is perhaps the most relevant case, where the government grants a patent to the developer of the substitute;

(*d*) competitive substitute, monopoly resource; if one views OPEC as acting as a collusive cartel, and the consuming governments as developing a substitute, the technology for producing which they will make freely available, this is an appropriate model for analyzing the oil market;

(*e*) one firm controls the natural resource, and another firm controls the substitute (duopoly).

II. The Basic Arbitrage Equation for a Competitively Owned Resource

Let S_t denote the stock of an exhaustible resource at t. We take it that the resource is owned competitively. Let p_t denote the spot price of the pre-extracted resource at t and let r_t denote the competitive market rate of interest. Consider a short interval of time $(t, t+\theta)$. Suppose that the probability that some specific event will occur during this interval is $\lambda_t\theta$. If the event does occur, the competitive price will be $\hat{p}_t$. We refer to $\hat{p}_t$ as the *fall-back* price of the resource at t. Presumably this fall-back price will depend on S_t, and the specific nature of the event. Thus, we write $\hat{p}_t = \hat{p}_t(S_t)$. If the event does not occur during this interval the price at $t+\theta$ will be $p_{t+\theta} = p_t + dp_t$. If speculators are risk neutral in this economy then in dynamic equilibrium one will have

$$\lambda_t\theta\hat{p}_t(S_t) + (1-\lambda_t\theta)(p_t+dp_t) = (1+r_t\theta)p_t, \tag{2.1}$$

which, on taking limits as $\theta \to 0$ yields the basic arbitrage condition

$$\frac{\dot{p}_t}{p_t} = r_t + \lambda_t\left[1 - \frac{\hat{p}(S_t)}{p_t}\right]. \tag{2.2}$$

---

[1] We follow the usual convention of defining the date of invention as the date at which it becomes possible to produce the substitute, the date of innovation as the date at which the substitute actually gets produced.

Equation (2.2) is, of course, a very general one; it will for instance be valid even when one contemplates an entire sequence of possible events over time. It represents the equilibrium condition at t, so that λ_t denotes the probability density that a specific event occurs at t conditional on it not having occurred earlier.

Certain special cases of (2.2) may now be mentioned. If either $\lambda_t = 0$ (i.e., there is no chance that the event will occur at t) or if $\dot{p}(S_t) = p_t$ (i.e., the event is a trivial one and has no bearing on the market for the resource), then $\dot{p}_t/p_t = r_t$. It is this special case that has been analyzed at length in the earlier literature.[1] Given that we have not specified the precise nature of the event and, consequently, that we do not know what the fall-back price is, it may appear as though equation (2.2) is far too general to enable one to obtain any specific insight into the rate of price change of the resource. But in fact, for one large class of cases the equation provides a useful inequality. Suppose the event contemplated at t is a "beneficial" one (e.g., the discovery of a new reserve of the resource or the invention of a substitute product). Then clearly $\dot{p}(S_t) \leqslant p_t$. In such a situation equation (2.2) implies that

$$r_t \leqslant \frac{\dot{p}_t}{p_t} \leqslant r_t + \lambda_t. \tag{2.3}$$

Thus under competitive equilibrium the possibility of a beneficial event occurring at t carries with it the implication that the rate of price rise of an exhaustible resource will be in excess of the rate of interest at t. But the rate of price rise will be bounded above by the sum of the interest rate and the probability density of the occurrence of the event at t conditional on its not having occurred prior to t. It is only when $\dot{p}(S_t) = 0$ (i.e., the event renders the existing stock worthless), that

$$\dot{p}_t/p_t = r_t + \lambda_t.[2]$$

As we have remarked earlier, in order to give more structure to the problem we shall visualize the possibility of the occurrence of a *single* event—the discovery of a substitute product that can be produced at unit cost $\bar{p}$. $\bar{p}$ is known with certainty but the date of occurrence of this invention is random. In other words, after the invention occurs (if it does occur, that is) there is no remaining uncertainty for the economy in question.[3] Given this, the information tree for our model is simple and is described, for discrete time, in Fig. 1. The world line $0A$ represents the non-occurrence of this single event.

---

[1] See, for instance, Hotelling (1931), Herfindahl (1967), Solow (1974a), Stiglitz (1977) and Sweeney (1974).

[2] This is the special case analyzed by Dasgupta & Heal (1977).

[3] It will become evident from the discussion below that if instead we considered the situation where at some date t it becomes known that the invention will occur at some subsequent date $t + T$, then there will be a discontinuity in the price of the resource at t, the date at which the information becomes available. There may be a further discontinuity in the price at $t + T$.

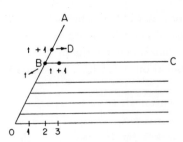

Fig. 1

The nodal point B represents the state of the world at t conditional on the discovery not having been made prior to t. At this point λ_t is the probability that the economy will follow the branch BC and $(1-\lambda_t)$ is the probability that it will move to D.

Now equation (2.2) has been arrived at from the postulate of risk neutral speculators. It is clear though that we could have as well arrived at it by postulating the existence of a complete set of contingent futures markets, thereby removing all uncertainty from the resource owners, provided that the market as a whole acts in a risk neutral manner to this particular risk (e.g., because its consequences are small relative to national income). Using continuous time representation and assuming λ constant and $\bar{p} = 0$ (i.e., the process is a Poisson one) $p_t e^{-(r+\lambda)t}$ would denote the price to be paid at the point 0 for the delivery of a unit of the resource at B.[1] It is also clear that equation (2.2) would describe the behavior of the imputed price of the resource, were both the resource and the substitute to be socially managed. This last is of course what basic welfare economics would imply.

In this paper our focus of attention will be on the rate of extraction and the fall-back price $\dot{p}(S_t)$, both of which will be endogenous for our system. Consequently, we take it that r_t is given exogenously and, for simplicity, that $r_t = r$ (a constant, >0). We shall also assume here that λ_t is uninfluenced by policy.[2] Again, for simplicity, we shall assume that $\lambda_t = \lambda$ (a constant).

III. Analysis of Socially Optimal Patterns of Allocation

We can use the arbitrage equation to solve for the equilibrium rate of extraction in the period prior to the invention occurring, provided we know the function $\dot{p}(S)$, giving the fall-back price as a function of the stock. We simplify as before and assume r and λ are independent of time. Our market is described then by the price differential equation,

---

[1] We shall naturally calculate the entire set of contingent prices after articulating the model in detail.
[2] In Dasgupta & Stiglitz (1980 *a* and *b*), Dasgupta, Gilbert & Stiglitz (1980 *a* and *b*) and Stiglitz, Gilbert & Dasgupta (1978) the rate of technical progress is taken to be endogenous.

$$\dot{p} = (r+\lambda)p - \lambda\hat{p}(S) \tag{3.1}$$

and by the extraction equation

$$\dot{S} = -D(p), \tag{3.2}$$

where $D(p)$ is the demand for the resource as a function of the price. We postulate that $D(p) > 0$ for all p. For simplicity we focus on the case of constant elasticity demand functions[1]

$$Q = p^{-\varepsilon}, \quad \varepsilon > 1.$$

In addition to these two differential equations, we need two boundary conditions: one gives the initial stock, S_0, and[2] the other says that, if the invention never occurs, we only use up the resource asymptotically.

$$\lim_{t\to\infty} S = 0, \quad S(t) > 0 \quad \text{for all finite } t. \tag{3.3}$$

In Fig. 2, we have presented the phase diagram for this market. Clearly, $\dot{p} = 0$ along the curve

$$p = \frac{\lambda\hat{p}(S)}{r+\lambda}$$

and above the $\dot{p} = 0$ curve, $\dot{p} > 0$.

In Stiglitz and Dasgupta (forthcoming) we establish the not surprising result that the fall back price is a declining function of the stock:

$$\hat{p}'(S) < 0.$$

Thus, the $\dot{p} = 0$ locus is downward sloping.

Since we postulate that $D(p) > 0$ for all p, the phase diagram appears as drawn. Working with the boundary condition (3.3) is slightly awkward; it is easier if we "guess" a value of p_0.

We can easily solve (3.1) and (3.2) for any given initial values of p_0, S_0. The solutions as a function of p_0 for given S_0 are illustrated in Fig. 3. The critical property of the solution[3]

---

[1] We choose this class of functions not only for the analytical simplicity, but also because we know that in the simplest models, it is precisely this class of functions which gives rise to no bias in the monopoly rate of resource extraction; Stiglitz (1977). The restriction to elasticities greater than unity is, of course, necessary in the analysis of monopolistic market structures.

[2] This is the familiar transversality condition of optimal growth theory. Clearly, if $\lim_{t\to\infty} S > 0$, the trajectory is inefficient, while if $S = 0$ at $t = T < \infty$, the "virtual" price, given our constant elasticity demand curve, is infinite for all $t > T$, and hence again the path cannot be efficient.

[3] It follows directly from observing that

$$\frac{\partial\dot{p}}{\partial p} > 0; \quad \frac{\partial\dot{p}}{\partial S} = -\lambda\hat{p}' > 0.$$

$$\frac{\partial\dot{S}}{\partial p} = -D' > 0.$$

$$\{p(t;\, S_0,\, p_0),\ S(t;\, S_0,\, p_0)\}$$

is that

$$\frac{\partial p}{\partial p_0} > 0, \quad \frac{\partial S}{\partial p_0} > 0, \quad \text{all } t,\ S_0;$$

at each t, price is a monotonically increasing function of initial price and the stock remaining is a monotonically increasing function of p_0.

Thus, for p_0 very large, $\lim_{t\to\infty} S_t > 0$,[1] while for p_0 very small, S becomes 0 in finite time. There thus exists a critical value of p_0, denoted p_0^s, satisfying the

---

[1] $\dfrac{\dot{p}}{p} = r + \lambda - \dfrac{\lambda \hat{p}}{p} < r + \lambda,$

i.e.

$$p < p_0\, e^{(r+\lambda)t}$$

$$-\dot{S} > p_0^{-\varepsilon}\, e^{-(r+\lambda)\varepsilon t}$$

$$S_t < S_0 - \frac{p_0^{-\varepsilon}(1 - e^{-(r+\lambda)\varepsilon t})}{(r+\lambda)\,\varepsilon} < 0$$

for

$$p_0 < \left[\frac{(r+\lambda)\,\varepsilon S_0}{1 - e^{-(r+\lambda)\varepsilon t}}\right]^{-1/\varepsilon}$$

Similarly

$$\frac{1}{\lambda}\frac{d\dot{p}/p}{dt} = -\frac{\hat{p}'\dot{S}}{p} + \frac{\dot{p}}{p}\frac{\hat{p}}{p}$$

It can be shown, as in Proposition 1 in Stiglitz & Dasgupta (forthcoming), that in the postinvention phase,

$$p = \hat{p}\, e^{rt}$$

$$\dot{S} = -D(p).$$

Hence

$$\frac{d\hat{p}}{dS} = -\frac{\hat{p}r}{D}, \quad \text{and}$$

$$\frac{1}{\lambda}\frac{d\dot{p}/p}{dt} = (r+\lambda)\frac{\hat{p}}{p} - \left(\frac{\hat{p}}{p}\right)^2 \lambda - r\frac{\hat{p}}{p} > 0$$

Hence

$$p > p_0\, e^{[r+\lambda - \lambda\hat{p}(S_0)/p_0]t}$$

Hence

$$\lim_{t\to\infty} S_t > S_0 - \frac{p_0^{-\varepsilon}(1 - e^{-(r+\lambda - \lambda\hat{p}(S_0)/p_0)\varepsilon t})}{(r+\lambda - \hat{p}(S_0)/p_0)\,\varepsilon} > 0$$

for p_0 sufficiently large.

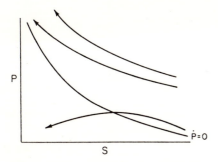

Fig. 2. Phase diagram for competitive economy.

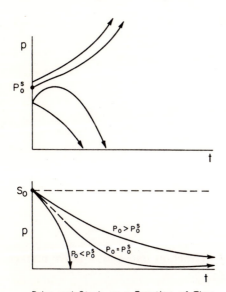

Price and Stock as a Function of Time

Fig. 3. Price and stock as a function of time

differential equations and the boundary conditions $S(0) = S_0$ and (3.3). It is also immediate that $dp_0/dS_0 < 0$.

We summarize the result as

Proposition 1. *There exists a unique, optimal extraction policy satisfying (3.1)–(3.3); the price of the resource increases monotonically over time at a rate less than $r + \lambda$; the price is a monotonically declining function of the size of the stock of the resource remaining:*

$$p^s = p^s(S) \frac{dp^s}{dS} < 0.$$

Proposition 1 provides a basis for comparing extraction rates with and without uncertainty; our intent here, however, is not in that comparison

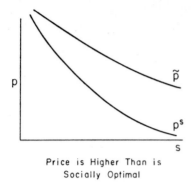

Price is Higher Than is
Socially Optimal

Fig. 4. Price is higher than is socially optimal.

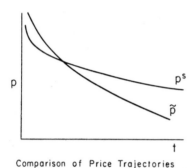

Comparison of Price Trajectories

Fig. 5. Comparison of price trajectories.

but in a comparison of the patterns of extraction under alternative market structures; see Dasgupta–Stiglitz (1979).

IV. Innovation Patented

The easiest comparison to make is between the socially optimal allocation and that where the substitute will be controlled by a single producer (e.g., as a result of a patent). This, of course, would typically be the case, except if the research was publicly funded; and the fact that the substitute is controlled by a single firm does not mean the economy is not competitive—in the conventional sense of that term—for there may have been competition in the research process, the patent holder being the winner in that competitive struggle, and there may be continuing competition for developing new, lower costs substitutes.

The analysis proceeds exactly as in the purely competitive (socially optimal) case, except the fall back price is now different. The fact that the market after the invention is not competitive leads the price after the invention to be higher than it is in the socially optimal allocation. Denoting by $\hat{p}$ the fall-back price in this market structure, it is shown in Stiglitz & Dasgupta (Proposition 3 b) that

$$\hat{\tilde{p}}(S) > \hat{p}^s(S) \tag{4.1}$$

To see what this implies for the pre-invention market, we need first to prove

Lemma 1. *Consider the following two pairs of differential equations:*

$$\dot{x} = ax - b_1(y) \quad b_1' < 0, \ b_1 > 0, \ a > 0 \tag{4.2a}$$

$$\dot{y} = -c(x) \qquad c' < 0, \ c > 0 \ \text{all} \ x > 0 \tag{4.2b}$$

$$x(0) = x_0 > 0 \tag{4.2c}$$

$$\lim_{t \to \infty} y = 0. \tag{4.2d}$$

$$\dot{x} = ax - b_2(y) \quad b_2'(y) < 0, \ b_2 > b_1 > 0 \tag{4.3a}$$

$$\dot{y} = -c(x) \tag{4.3b}$$

$$x(0) = x_0 > 0 \tag{4.3c}$$

$$\lim_{t \to \infty} y = 0. \tag{4.3d}$$

Let $x_1(t)$, $y_1(t)$ be the solution to (4.2), $x_2(t)$, $y_2(t)$ be the solution to (4.3) and let $x_1(y)$ and $x_2(y)$ be the solution in (x, y) phase space. Then

$$x_1(y) < x_2(y)$$

for all y.

Proof. Assume $x_1(\tilde{y}) \geqslant x_2(\tilde{y})$ for some $\tilde{y}$. Let $y_1(t_1) = \tilde{y}$, $y_2(t_2) = \tilde{y}$. Then when $y = \tilde{y}$, $\dot{x}_1 > \dot{x}_2$, $\dot{y}_1 \geqslant \dot{y}_2$. Since

$$\frac{\partial \dot{x}}{\partial x} = a > 0, \quad \frac{\partial \dot{x}}{\partial y} = -b' > 0,$$

$x_1(t+t_1) > x_2(t+t_2)$, $y_1(t+t_1) > y_2(t+t_2)$. Hence, (4.2d) and (4.3d) cannot both be satisfied.

An immediate implication of Lemma 1 and inequality (4.1) is

Proposition 2. *Prior to invention, the economy in which the substitute will, upon discovery, be controlled by a patent, is excessively conservationist:*

$$\tilde{p}(S) > p^s(S)$$

for all S.

In Stiglitz & Dasgupta, it was shown (Proposition 3 b) that if the substitute were controlled by a monopolist, then the date of innovation, for any given initial stock of the resource at the date of invention, would be delayed. *The actions of the (competitive) resource owners prior to invention thus exacerbate further this delay in the date of innovation.*

V. The Monopoly Arbitrage Equation

The monopolist, in deciding his extraction policy, compares the marginal return from extracting now, to the expected marginal return from postponing. Let $V^m(S)$ be the present discounted value of a stock S for a pure monopolist (i.e., a monopolist in both the resource and the substitute). Thus

$$V^m(S) \equiv \max_{\{Q(t),\,T^m\}} \left\{ \int_0^{T^m} R(Q)\,e^{-rt}dt + e^{-rT^m} \max \frac{R(Q) - \bar{p}Q}{r} \right\} \text{ s.t. } \int_0^{T^m} Q(t)\,dt \leqslant S. \tag{5.1}$$

where

$$R(Q) = p(Q)Q = \text{revenues when sales are } Q$$
$$T^m = \text{date of exhaustion of the natural resource}$$

$\max \dfrac{R(Q) - \bar{p}Q}{r}$ = the present discounted value of the profits generated by the substitute.

(We know from Stiglitz & Dasgupta (forthcoming) Proposition 2a, that the optimal pattern of production and resource extraction for a pure monopolist always takes on the form represented by (5.1), i.e., a first phase in which only the resource is extracted; and a second phase in which only the substitute is produced.)

Let $\hat{M}(S)$ represent the *fall-back* marginal revenue, the marginal return to having an extra unit of the resource at the date of invention, i.e.,

$$\hat{M}(S) \equiv V^{m\prime}(S) \tag{5.2}$$

Thus, by reasoning identical to that employed in Section II, the expected present discounted value of postponing extraction from t to $t+\theta$ is

$$\lambda_t \theta \hat{M}(S_t) + (1 - \lambda_t \theta)(M + dM) = (1 + r_t \theta) M \tag{5.3}$$

which on taking limits as $\theta \to 0$, yields the basic monopoly arbitrage condition

$$\frac{\dot{M}_t}{M_t} = r_t + \lambda_t \left[1 - \frac{\hat{M}(S_t)}{M_t} \right] \tag{5.4}$$

With constant elasticity demand curves, this then can be written as

$$\frac{\dot{p}}{p} = r_t + \lambda_t \left[1 - \frac{\hat{M}(S_t)}{\dfrac{\varepsilon - 1}{\varepsilon} p} \right] \tag{5.5}$$

Thus, a comparison of the competitive and monopoly pre-invention rates of extraction is reduced to a comparison between

$$\frac{\hat{M}(S)\,\varepsilon}{\varepsilon - 1} \quad \text{and} \quad \hat{p}^s(S) \tag{5.6}$$

For a pure monopolist, the marginal revenue from resource extraction is (in present discounted value terms) identical at all dates after this invention and equal to

$$\hat{M}(S) = \hat{p}^m(S)\frac{\varepsilon - 1}{\varepsilon}$$

where $\hat{p}^m$ is the level to which the monopolist drops the price after the invention. Since we established in Stiglitz and Dasgupta (forthcoming) that (Proposition 2 b)

$$\hat{p}^m(S) > p^s(S) \quad \text{for all } S$$

it immediately follows (upon again using Lemma 1)

Proposition 3 a. *During the pre-invention phase the monopolist is unambiguously more conservation-minded than is socially optimal.*

By the same token, from Lemma 1 and the fact that $\tilde{p}(S) < p^m(S)$ for all S (Stiglitz & Dasgupta, Proposition 3 b), the pure monopoly price exceeds the price when the resource is competitively controlled but the substitute is controlled by a monopolist, we immediately observe

Proposition 3 b. *An economy in which the resource is competitively owned but the substitute is controlled by a patent is more conservation-minded than is socially optimal but less so than a pure monopolist.*

It is this limited competition case which conforms to our intuition that mixed cases (limited competition) ought to lie between the polar cases of pure competition and pure monopoly.

VI. Resource Owned by a Monopolist, Substitute Competitively Produced

The analysis of this case is almost identical to that of the preceding except that now we define

$$V^* \equiv \max \int_0^\infty R(Q)\,e^{-rt}dt \tag{6.1}$$

s.t. $p(t) \leqslant \bar{p}$

and

$$\int_0^\infty Q(t)\,dt \leqslant S_0$$

V^* is the maximized present discounted value of the revenue generated by the resource when there is a competitive substitute available, which will be produced at the price $\bar{p}$. (The asterisk is simply used to denote this market structure.)

We can again show (with precisely the same line of argument) that, prior to the invention, the resource-monopolist will allocate his resources so that the. arbitrage equation (5.4) is satisfied; but now, $\hat{M}(S)$, the marginal revenue of an extra unit of resource in the post-invention era, must be calculated in a different way. The marginal revenue is no longer necessarily proportional to price. In our earlier analysis, we showed that if T_2 denotes the date of final exhaustion of the natural resource, measured from the date of invention, then the marginal revenue at the date of invention is

$$\bar{p}e^{-rT_2}$$

and that this equalled $\hat{p}(\varepsilon-1)/\varepsilon$ if and only if the fall-back price, $\hat{p}<\bar{p}$, i.e. if and only if S were large. The competitive supply of the substitute imposes a limit, $\bar{p}$, on the price which the monopolist of the resource can charge; extra units of the resource are sold at $\bar{p}$ at T_2.

To analyze prices prior to invention, we thus need to compare $(\hat{M}^*(S)\varepsilon)/(\varepsilon-1)$, marginal revenue in this regime, with $(\hat{M}^m(S)\varepsilon)/(\varepsilon-1)$ under pure monopoly, on the one hand, and with p^s and $\tilde{p}$ on the other. Proposition 4b of Stiglitz & Dasgupta established that

$$\hat{M}^m(S) < \hat{M}^*(S) \tag{6.2}$$

even when $\hat{p}^m > p^*(S)$, i.e. even when the monopoly price exceeded the price in this regime, marginal revenue was less, while Propositions (2b) and (3b) established that

$$\frac{\hat{M}^*(\varepsilon-1)}{\varepsilon} > \hat{p}^m > p^s.$$

Using (6.2) and Lemma 1, we thus obtain

Proposition 4. $p^*(S)>p^m(S)\tilde{p}>(S)>p^s(S)$. *If the substitute is supplied competitively while the resource itself is controlled by a monopolist, then pre-invention price is even higher than in pure monopoly, and hence innovation is delayed even more.*

The intuitive interpretation of this result is that, by pursuing a more conservation-minded strategy, the resource owner is able to extend the period of his "monopoly" control of the market. This result is strengthened when the rate of technical progress is endogenous; see Dasgupta, Gilbert & Stiglitz (1980*b*).

VII. Duopoly

The final case we consider is that where the resource and the substitute are controlled by two different monopolists. We then have, after the invention, a (potential) duopoly market. In our earlier study, we were able to characterize the Nash equilibrium as having three phases: in the first only the resource owner produced, and price was lower than $\bar{p}$; in the second both are produced, and market price is between $\bar{p}$ and $\bar{p}(\varepsilon/(\varepsilon-1))$; and in the final stage, $p = \bar{p}(\varepsilon/(\varepsilon-1))$ and only the substitute is produced. We were able to obtain a differential equation describing the price movements in this duopoly model, and to show that $\hat{p}^d(S) \gtrless p^m(S)$ as $S \gtrless S^*$: the duopoly fall-back price, $\hat{p}^d(S)$, was smaller or larger than the pure monopoly price as the stock at the date of invention was smaller or larger than some critical level S^*.

But just as in the previous case, the relationship between price and marginal revenue is complicated. In Stiglitz & Dasgupta (forthcoming), we were able to show that

$$\hat{M}^d(S) > M^m(S) \quad \text{for} \quad S < \hat{S} < S^* \text{ and } S \geqslant S^*; \tag{7.1}$$

even though, for $S < \hat{S}$, the duopoly price is higher, marginal revenue is lower.[1] Indeed, since

$$\hat{M}^d = p(1 - \mu/\varepsilon)$$

where μ is the fraction of the market supplied by the resource producer, for S near 0, $\mu \approx 0$, and hence

$$\hat{M}^d \approx p = \frac{\bar{p}\varepsilon}{\varepsilon - 1},$$

which may be considerably greater than $\hat{M}^m$:

$$\hat{M}^m \approx \bar{p}.$$

We can use these results, in conjunction with Lemma 1, to establish

Proposition 5. *In the Nash-equilibrium duopoly, the resource owner in the pre-invention era is more conservation minded than the pure monopolist and (a fortiori) than is socially optimal, provided the initial stock of the resource is not too large: prices, at least initially, with limited competition may be much higher than with pure monopoly.*

VIII. Concluding Remarks

The basic results of this paper can be summarized in Fig. 6, giving the price of the resource as a function of the stock available, during the period prior to the invention having been discovered.

---

[1] For $\hat{S} < S < S^*$, we conjecture that (7.1) still holds, but we have not been able to prove it.

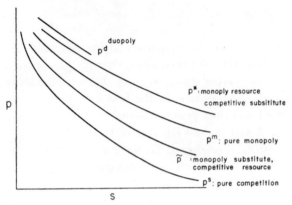

Fig. 6.

The result of particular interest relates to the competitive provision of the substitute: one of the strategies proposed for responding to the OPEC cartel is for the consuming nations to develop substitutes which they could then ensure would be competitively supplied. The response of a resource monopolist to this change in market structure would be to *raise* his prices, not to lower them. Although in the post-invention phase, his prices would be lower, his marginal revenue might be higher, and it is the effect on marginal revenue which is critical for determining his supply behavior in the pre-invention phase.

If one is concerned with using competition as a mechanism for lowering the price of a natural resource, but if one is unable to attain a perfectly competitive market, it appears to be critical precisely how competition is introduced: some forms of competition may serve to raise rather than to lower prices.

References

Arrow, K.: Economic welfare and the allocation of resources for inventors. In *The rate and direction of inventive activity: Economic and social factors* (NBER) (ed. R. Nelson). Princeton Univ. Press, 1962.

Dasgupta, P., Gilbert, R. & Stiglitz, J. E.: Invention and innovation under alternative market structures: The case of natural resources. Econometric Research Program Memorandum No. 263, Princeton, March 1980a.

Dasgupta, P., Gilbert, R. & Stiglitz, J. E.: Energy resources and research and development. Erschöpfbare Resourcen. Duncker and Humbolt, Berlin, 1980b.

Dasgupta, P. & Heal, G. M.: The optimal depletion of exhaustible resources. *Review of economic Studies*, Symposium on the Economics of Exhaustible Resources, pp. 3–28.

Dasgupta, P. & Heal, G. M.: *Economic theory and exhaustible resources*. Chapter 11. (Cambridge Handbook), 1978.

Dasgupta, P. & Stiglitz, J. E.: Industrial structure and the nature of innovative activity. *Economic Journal*, June 1980a (*90*), pp. 266–293.

Dasgupta, P. and Stiglitz, J. E.: Uncertainty, Industrial Structure and the Speed of R & D, *Bell Journal of Eco-*

nomics, Vol. 11, No. 1, Spring 1980*b*, pp. 1–28.

Dasgupta, P. & Stiglitz, J. E.: Resource depletion under technological uncertainty. *Econometrica*, Jan. 1981, Vol. 49, pp. 85–104.

Herfindahl, O. C.: Depletion and economic theory. In *Extractive resources and taxation* (ed. M. Gaffney). University of Wisconsin Press, Madison, 1967.

Hotelling, H.: The economics of exhaustible resources. *Journal of Political Economy 39*, 137–175.

Solow, R.: Intergenerational equity and exhaustible resources. *Review of Economic Studies*, pp. 29–45, 1974*a*.

Solow, R. M.: The economics of resources or the resources of economics. *American Economic Review*, Papers and Proceedings, pp. 1–14. May 1974*b*.

Stiglitz, J. E.: Growth with exhaustible natural resources: Efficient and optimal growth paths. *Review of Economic Studies*, Symposium on the Economics of Exhaustible Resources, pp. 123–138, March 1974*a*.

Stiglitz, J. E.: Growth with exhaustible natural resources: Competitive growth paths. *Review of Economic Studies*, pp. 139–152, March 1974*b*.

Stiglitz, J. E.: Monopoly and the rate of extraction of exhaustible resources. *American Economic Review*, 1977.

Stiglitz, J. E. & Dasgupta, P.: Market structure and resource depletion. *Journal of Economic Theory*, forthcoming.

Stiglitz, J. E., Gilbert, R. & Dasgupta, P.: Invention and innovation under alternative market structures: The case of natural resources. Mimeographed, 1978.

Sweeney, J.: Economics of depletable resources: Market forces and intertemporal bias. *Review of Economic Studies 44*, 125–142, February 1974.

SCARCITY, EFFICIENCY AND DISEQUILIBRIUM IN RESOURCE MARKETS

*Geoffrey Heal**

University of Essex, Colchester, England

Abstract

Possible interpretations of the concept of scarcity in extractive resource markets are presented and then related to traditional economic ideas of efficient patterns of resource use. Sources of inefficiency in resource markets are then considered, with the main emphasis being on disequilibrium. Various simple models of disequilibrium suggest that resource markets are more stable if traders' expectations are based on quantity rather than price information, so that the provision of such information increases efficiency.

I. Introduction

I need hardly emphasise the importance of the issue of resource scarcity, and the greater part of my paper will be devoted to a discussion of what might be meant by this term, and to an analysis of the relationships between scarcity and prices under various conditions. In these introductory remarks, I would just like to point out that although the issue of resource scarcity arises very naturally and seems self-evidently important, there are underlying it some conceptual issues which are, in fact, very elusive. Other writers, such as Fisher (1979) and V. K. Smith (1980), have recently noted this point,[1] but my approach will be rather different and perhaps more sceptical of the value of scarcity as a concept. The fundamental issue, of course, is: what does one mean by saying that an exhaustible resource is scarce? According to the Oxford English Dictionary, scarce means "restricted in quantity, size or amount". In this literal sense, then, an exhaustible resource, of which by definition only a finite amount is available, is always scarce. The question "Are such resources scarce?" becomes a rather trivial one. This suggests, of course, that the issue that really concerns us is not resource scarcity *per se*, but its economic importance. (Smith refers to this as adequacy.) In order to discuss this, we need

---

* I am grateful to Graciela Chichilnisky and Partha Dasgupta for valuable comments.
[1] The detailed references are to Anthony C. Fisher, "Measures of natural resource scarcity" in *Scarcity and Growth Reconsidered*, ed. V. Kerry Smith, John Hopkins Press, 1979, and V. Kerry Smith, "The evaluation of natural resource adequacy: Elusive quest or frontier of economic analysis" in *Land Economics*, 1980.

a measure of scarcity, or of its economic impact. What is at issue is whether the resource is scarce relative to the demands we would like to make on it. This sounds suspiciously like the traditional question of the adequacy of supply relative to demand, which is itself not altogether well-formulated. In a market system in equilibrium, supply will *always* equal demand, simply because the price adjusts to bring the two into line. Low supplies mean high prices, and vice versa. And high prices mean that the use of the resource is confined to high-priority, high-productivity activities, so that in consequence the productivity to society of the acquisition of more of the resource—its shadow price, in technical terms—is great.

I hope that this has illustrated the point that underlying the apparently straightforward issue of resource scarcity, are some elusive concepts. Certainly we are not just interested in asking whether exhaustible resources are scarce, because, according to the dictionary, this issue is trivial. And in a market system we cannot ask whether supply is sufficient to meet demand, because the price mechanism will take care of that. What question are we then left with, if any? I maintain that we are still left with an important and difficult question, and one that any economist will instantly recognise as central to the discipline. We are in fact left with the question: "Are we making the best use of our scarce resources?". But before turning to develop this issue further, I want to spend some time looking at a famous empirical study, that of Barnett and Morse.[1] I want to do this because their work illustrates beautifully the importance of choosing most carefully between different formulations of the question. If one poses the question one particular way, as Barnett and Morse did, an optimistic answer seems appropriate. But posing it in the way I prefer leads one to draw more pessimistic conclusions from their work. So it is not intellectual hairsplitting to insist that in this area, the question should be posed with great care.

II. Prices in the Long Run

Barnett & Morse, as is well known, analysed the scarcity of natural resources in the U.S.A. over the period 1870 to 1953. They did this by compiling time series on prices and costs for various groups of commodities—minerals, forest products, agricultural products, etc. My concern here is limited to exhaustible resources, and therefore to their findings on minerals. In essence, Barnett and Morse found that for both mineral prices and mineral costs, the trend was decreasing, or at least non-increasing. A basic assumption of theirs, discussed at some length in their book, was that prices and costs were appropriate measures of scarcity, and they therefore deduced from their findings that mineral resources had become less scarce in the U.S.A. over the period 1870 to 1953.

---

[1] Harold J. Barnett and Chandler Morse, *Scarcity and Growth*. Johns Hopkins Press, 1963.

For the moment, I merely want to express some slight scepticism about their identification of prices and costs with scarcity. It is an intuitively appealing identification—until one asks oneself whether oil was really three times as scarce in 1974 as in 1973. Obviously, then, this is an issue to which we shall have to return. But in the meantime it is useful to look at their empirical work, and related work, in more detail. There are several criticisms that can be made of the Barnett-Morse work—as, indeed, is the case of almost any empirical work. One point that has been developed by Kerry Smith[1] is that their cost index is for various reasons unsatisfactory. A further point is that they present only an index of prices for all minerals: the aggregation involved in this could well hide divergent trends. Finally, their analysis of the data is purely graphical and visual, and so could easily miss subtle changes in the nature of the price path.

Fortunately, there have been a number of interesting recent attempts to correct these deficiencies. The work by Potter and Christy and Manthy has made available to us updated price series for the period 1870 to 1973, and the statistical analyses of Kerry Smith have confirmed the absence of a clear linear trend in the aggregate price data. Work by my colleague Barrow,[2] using price series for individual exhaustible resources over the extended preiod 1870 to 1973, suggests that, while there is no clear linear trend, there is in fact a clear quadratic trend. Until the end of the interwar period, most of these series show a significant downward trend and, from then on, a slight but significant upward trend. This finding of a rather shallow U-shaped pattern in the long-run price movement has considerable theoretical implications, to which I shall return in due course.

Let me now summarise my reference to this empirical work. On the basis of longer-term and more disaggregated data, the original conclusion of a downward long-term trend in prices must be queried. And the original Barnett-Morse implication that lower prices mean less scarcity, must also be queried. Recent behaviour of oil prices clearly does not permit one to accept such a generalisation, and indeed, it could be argued that low prices lead to high consumption rates, rapid depletion and low remaining stocks. On this interpretation, then, low prices, rather than being a sign of abundance, are in fact a cause of scarcity. It might be noted that this is the kind of argument used at one point by the Iranians to justify an increase in the price of oil, and I shall argue later that it has some force. For the moment, I would just like to note that accepting price as a measure of scarcity leads one to draw optimistic conclusions from the long-term price data, whereas interpretation of low prices as a possible cause of scarcity leads of course to quite the opposite frame of mind.

---

[1] See the paper by V. Kerry Smith in *Scarcity and Growth Reconsidered*, op. cit.
[2] Michael Barrow, M.A. Thesis, University of Sussex.

III. Scarcity and Efficiency

I asserted earlier that questions such as "are resources scarce?" or "will their supply meet demand?" are simply not well-posed, and that the only sensible question is: "are resources being used efficiently?".[1] Certainly, this last is a recognisable and indeed conventional question, to which there is a non-trivial answer, so that this way of formulating the question has intellectual merits. It also has practical and operational merits. If resources are being used inefficiently, in the standard economic sense, then there is, so to speak, slack in the system and more economic value could be obtained from a given resource input. This should alert us to the possibility of a beneficial rearrangement of the resource-use system.

My contention, then, is that the only well-defined and operational question is: "are our exhaustible resources being used efficiently?". Of course, the question, as posed, comes naturally to economists, and we can turn to an established body of welfare economics and resource allocation theory in an attempt to find the constituents of an answer. In particular, we know from this literature that, under certain circumstances,[2] competitive markets will allocate resources efficiently. I shall therefore proceed from this literature as a base, and ask two questions:

(1) How would one recognise an efficient allocation of exhaustible resources? What data enables one to test the efficiency of a use pattern?
(2) What factors might cause departures from an efficient pattern of use, and what types of departures might these be?

Once I have answered these questions, I shall return to such empirical evidence as is available, to see if it helps us in deciding whether the allocation to date has been efficient, or whether there have been some recognisable causes of inefficiency. So ultimately my answer to the question originally posed—how do we measure resource scarcity—is that we don't. We look instead for inefficiency in resource use, and design policy measures to reduce it.

IV. Prices in Efficient Markets

I turn now to the question of how one would recognise whether resources are being used efficiently. Since Hotelling's work in the 1930s,[3] it has been re-

---

[1] I have in mind the standard microeconomic concept of Pareto efficiency: the state of the economy is efficient if it is impossible make one person better off without making another worse off.

[2] These conditions are clearly set out in, for example, T. C. Koopmans, "The price system and the allocation of resources", in T. C. Koopmans, *Three Essays on the State of Economic Science*. See also chapter two of P. S. Dasgupta and G. M. Heal, *Economic Theory and Exhaustible Resources*, Cambridge Economic Handbooks, 1979, for a discussion related to the present context.

[3] H. Hotelling, "The economics of exhaustible resources", *Quarterly Journal of Economics*, 1931.

cognised that, under certain conditions, a necessary (though not sufficient) condition for efficient use of a resource is that the difference between its price and marginal extraction cost should rise over time at the rate of interest. Formally,

$$(p_t - MEC_t) = (p_0 - MEC_0) \exp{(rt)}, \tag{1}$$

where p_t, MEC_t are price and marginal extraction cost at date t, and r is the interest rate. This may be rewritten as

$$\frac{\dot{p}}{p} = r \frac{(p-C)}{p} + \frac{\dot{C}}{C} \cdot \frac{C}{p}, \tag{2}$$

where p and r are as before, C is the marginal cost of extraction, and all variables are functions of time. Equation (2) has a very straightforward interpretation, and states that the price should rise at a rate which is a weighted average of the interest rate and the rate of change of extraction costs, with the weights on the interest rate and the cost changes being respectively the contributions of royalty and cost to price. Some discussion of the implications of equation (2) is in order. Clearly the first term on the right is positive, whereas the second will be negative if costs are falling and positive if they are rising. One might, of course, expect that costs would fall over an initial period during which development of the resource is being opened up, and during which there is technical progress, learning, etc., but that, in the long-run, diminishing returns in the form of the exhaustion of high-grade deposits would force costs up. It is also clear that one would initially expect the royalty element in prices to be low, and hence the first term on the right of (2) to be small. We are therefore drawn to the conclusion that initially $\dot{p}/p$ is likely to be negative, because the second term in (2) is negative and the first small, but that in the long-run $\dot{p}/p$ would become positive. In an efficient market, prices thus fall initially, but rise in the longer term, generating a U-shaped path.

You will recall that, when referring earlier to extensions of the Barnett–Morse work by my colleague Barrow, I mentioned that he had discovered the time-series for a number of resource prices over the period 1870 to 1973 to show a clear quadratic or U-shaped trend. You can now see the significance of his finding: it is qualitatively similar to the predictions of (2), and to the characteristics that would be observed if resource-use satisfied the conditions necessary for efficiency. Of course, one should not leap to optimistic conclusions from this: many factors have been omitted from both the theory and the empirical work. However, the coincidence is suggestive.

V. Inefficiency: Imperfect Compatition

I have now discussed the sort of price movements that would be generated by efficient, competitive markets, and noted with some interest that their salient features seem to resemble those of observed data. Clearly this similarity must be treated with great caution, if only because many of the relevant markets are evidently not competitive. It is obviously necessary to characterise the price paths that would result from imperfectly competitive markets, to analyse the resulting inefficiencies, and to enquire whether these paths might not also resemble observed data. An interesting outcome of this investigation will, in fact, be that in certain cases, imperfect competition in resource markets will actually cause little or no departure from the efficient competitive outcome.

V.1. *Monopoly*

Suppose now that a monopolist owns the entire stock S_0 of an exhaustible resource, and faces a demand curve $p\{R(t)\}$, where $R(t)$ is the rate at which the resource is supplied to the market at date t. Then, neglecting extraction costs, he will wish to choose a supply profile $R(t)$ so as to maximise

$$\int_0^\infty R(t) \, p\{R(t)\} \, e^{-rt} \, dt$$

subject, of course, to the constraint that

$$\int_0^\infty R(t) \, dt = S_0.$$

Letting $M(t)$ stand for marginal revenue, and if $\eta\{R(t)\} \leqslant 0$ is the elasticity of demand at supply $R(t)$, then it is a simple matter to verify that the solution must satisfy

$$\frac{\dot{p}(t)}{p(t)} = r - \frac{\dot{\gamma}(t)}{\gamma(t)}, \tag{3}$$

where $\gamma\{R(t)\} = 1 + 1/\eta(R)$. A comparison with the equivalent competitive case in (2) is now simple: as extraction costs are zero, the difference between the price paths depends solely on $\dot{\gamma}(t)/\gamma(t)$. A very simple case is that of a constant-elasticity demand function: in such a case, $\dot{\gamma}(t) = 0$ and the monopolistic and competitive price paths are identical in all respects. Such a case is clearly rather striking.

  More generally, of course, there are differences, and these depend on the nature of the demand function. Clearly

$$\dot{\gamma}(t) = -\frac{d\eta}{dR} \cdot \frac{\dot{R}}{\eta^2}$$

and hence

$$\text{Sign } \dot{\gamma} = \text{Sign } \frac{d\gamma}{dR}.$$

Consequently:

$$
\left.
\begin{aligned}
&\text{if } \frac{d\eta}{dR} > 0, \frac{\dot{p}}{p} < r \\[2mm]
&\text{if } \frac{d\eta}{dR} < 0, \frac{\dot{p}}{p} > r
\end{aligned}
\right\}
\tag{4}
$$

and it appears that the behaviour of the price will depend upon whether the elasticity of demand increases or decreases with output. Either situation seems possible—for example, one could argue that, as the price of a resource such as oil is lowered, it cuts into markets which had traditionally been the preserve of other fuels. In such cases, oil's advantage would be marginal and easily lost by small price changes—implying that lower prices and higher outputs lead to larger absolute demand elasticities and lower values of η. The contrary is equally plausible: it might be true that as the price is raised, this increases the incentive to invent substitutes that did not previously exist, or to speed up work on alternatives whose development had been held in abayance while the resource price was low. In such a case, increasing price and lowering output would reduce η. This kind of issue is analysed in detail by Chichilnisky,[1] who derives information about the elasticity of demand for oil from an analysis of the general equilibrium impact of oil price changes on the importing country. She shows that this elasticity may depend in an important way on the internal configuration of the economy, and so can be expected to vary over time.

V.2. Oligopoly

Oligopoly is usually difficult to analyse, and one would expect the imposition of dynamics, so essential for the study of resource depletion, to compound this difficulty. Interestingly, this is not so: if there are N identical and competing sellers of a resource, then one can show[2] that in an intertemporal Nash equilibrium (where each takes as given the entire price profile of each of his rivals) the price once again moves according to

$$
\frac{\dot{p}(t)}{p(t)} = r - \frac{\dot{\gamma}(t)}{\gamma(t)},
\tag{5}
$$

where $\gamma(t)$ is now

$$
\gamma = N + 1/\eta(R).
$$

Qualitatively, the price path relates to the competitive case exactly as in the case of monopoly, though the differences are smaller and tend to zero as N increases without limit.

---

[1] G. Chichilnisky, "Oil supplies, industrial output and prices: A simple general equilibrium macro analysis," Essex Economics Paper, Jan. 1981.
[2] This result is established in Dasgupta and Heal, op. cit.

V.3. *Asymmetrically Placed Oligopolists*

For many resources, the assumption that oligopolists are identical is unaccept-
able. What one observes in many markets (such as oil or bauxite) is the presence
of a dominant supplier or group of suppliers (typified by OPEC in the case of
oil, and referred to hereafter as the cartel) and a group of smaller and relatively
independent producers, who we shall refer to as the competitive fringe. In
such a situation, Nash's equilibrium concept is probably not appropriate: one
would expect the cartel to announce its policies having taken into account the
response of the competitive fringe, with this latter then responding passively.
An equilibrium concept which describes such a situation well is that of von
Stackelberg, and I shall analyse such an equilibrium in some simple cases.

Suppose that a resource can be extracted at negligible cost, and that there
is a cartel consisting of all resource owners. Suppose further that there is a
perfect substitute for the resource that can be produced at a price $\bar{p}$—I shall
refer to this as the backstop technology. Many producers have access to this
backstop technology, and they collectively form the competitive fringe.

The market equilibrium that results in such a situation is easily discovered.
Clearly the resource price will never exceed $\bar{p}$: equally clearly, if there is an
interval of time over which the price is less than $\bar{p}$, the cartel will wish to ensure
that equation (4) holds and the present value of marginal revenue rises at the
rate of interest. It would thus be reasonable to expect a period during which
prices rise in such a way as to keep present value marginal revenue constant,
followed by a period during which the price is infinitesimally less than $\bar{p}$, and
it can indeed be shown formally that a market equilibrium has this character.[1]
Of course, in the case of an isolelastic demand function, matters are somewhat
simpler: if the elasticity exceeds one in absolute value, then during the initial
interval prices will rise at the rate of interest (see the discussion following (5)).
If the elasticity is less than unity in absolute value, the cartel will of course
price just below $\bar{p}$ as long as it supplies the resource.

It is obvious that equilibria of the type we have just described involve a
form of limit pricing: the cartel prices to keep the competitive fringe and the
backstop technology out of the market. This has a very important implication
for the conduct of technical change designed to lower the cost $\bar{p}$ of the backstop. The point is that a reduction in $\bar{p}$ will just result in an equal reduction
in the cartel's price, with the substitute still being priced out of the market.
The private return to such cost reducing technical change will therefore be
zero. The social return, however, may be very great—for the market price of
the resource will have been forced down by the amount of the cost reduction,
with a corresponding increase in consumer surplus generated.

---

[1] Again, Dasgupta and Heal, op. cit. gives details.

V.4. *Imperfect Competition: Conclusions*

It is as usual difficult to generalise about the effects of imperfect competition. However, we have seen the rather interesting proposition that, under certain circumstances, monopoly and oligopoly will lead to little or no departure from the competitive outcome, and hence little or no inefficiency. In general, the nature of the inefficiency caused by imperfect competition will depend crucially on the nature of the demand function, and there is simply not enough information about these demand conditions for one to hazard a sensible guess. Of course, most demand functions estimated econometrically display constant elasticity, so that if we believed these to be accurate estimates, we could be optimistic. But one could argue that the imposition of a constant elasticity is an unjustifiable restriction: unfortunately it is not easy to proceed in any other fashion.

It seems that, on balance, imperfect competition cannot be said to be clearly and on purely theoretical grounds likely to be a major cause of efficiency losses, and this suggests that our interest in possible inefficiency should lead us to look at other issues, issues which are perhaps more clearly connected with the dynamic structure of the problem.

VI. Disequilibrium and Limited Information

The models of market behaviour discussed so far have all rested on two major assumptions, assumptions which have been more implicit than explicit. Let me now focus sharply on these. One assumption is an assumption of market equilibrium: it is a condition of market equilibrium that the return to holding a resource should equal the return to holding alternative assets, and it is this latter return that is symbolised by the interest rate. All of our earlier discussions have taken the form of deducing the implications of this equality of returns under various market structures. It is now time to admit the possibility that equilibrating forces may not be sufficiently strong to equalise these returns, and to enquire into the consequences of this.

Another assumption more implicit than explicit in the earlier discussions, was that traders always have accurate expectations, or full information, about future prices. Clearly this is a considerable over-simplification: people form expectations about market trends which are often wrong, and then revise their expectations in a manner which will be affected by many factors, including possibly the differences between actual and expected outcomes. Clearly we must analyse the likely outcome in markets where this type of phenomenon occurs, and enquire again whether it is liable to lead to inefficiency.

In view of these observations, I shall now discuss briefly models of markets where disequilibrium, in the sense of discrepancies in the rates of return on different assets, is a possibility, and where information about the future is imperfect and traders are guided by expectations. I consider first of all, and in rather general terms, the disequilibrium aspect of the problem.

The rates of return on different assets may become unequal because for example, a major technological development changes the profitability of capital or the productivity of resources, or because of some unexpected shock to the economy. Given this possibility, it is obviously important to enquire whether such a discrepancy is likely to be self-correcting. The next section presents a series of models which are concerned with these issues. Figure 1 illustrates in summary form some of the main feedback effects that are relevant.

VI. 1. *Elementary Models of Disequilibrium Behaviour*

It is clear from the above that the problems with which we are concerned are sufficiently complex, that we are unlikely to be able to construct a single model which incorporates all of the forces at work. It is also clear that how traders form expectations about future prices will be very important, and we begin by developing this point.

Consider, for example, the market's response to the observed depletion of a resource stock. It may become apparent to dealers that it will not always be possible to meet current levels of demand, and rumours of future shortages will cause prices to rise. This price increase, if it is expected to continue, will cause sellers to hold stocks off the market, and buyers to economise on their use of the commodity and start searching for substitutes. These two moves reduce future demand and raise future supply, thus tending to prevent the development of the anticipated shortage.

What we have just described is obviously an ideal, and many things could go wrong. For example, sellers might not expect the initial price increase to continue, so that an initial price increase consequent upon rumours of future shortages might actually bring a wave of selling designed to take advantage of what are thought to be temporarily high prices. This would, of course, worsen the possible market imbalances.

Obviously, which of these outcomes occurs will depend on how sellers form their expectations of future prices: if they do this by extrapolating current rates of change, the increase will have the desired effect of encouraging them to hold stocks off the market. If on the other hand, they have in mind some idea of a normal price, perhaps a weighted average of past prices, then they will regard a sharp rise as a temporary phenomenon to be taken advantage of by selling. Their willingness to bear risks will also come into the picture. If, as seems likely, they are not certain whether the upward trend in prices will continue, then the more risk-averse may well cash in on present high prices by selling, while gamblers keep their goods off the market.

In order to examine some of these possibilities more closely, I set up and comment briefly on some simple models of disequilibrium behaviour in a resource market. I would emphasise that these are very preliminary ventures

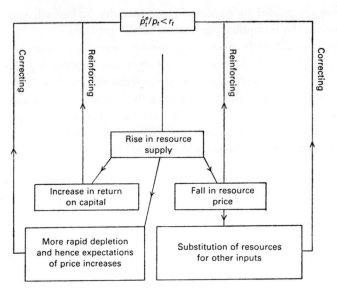

Fig. 1.

into a field in which a great deal clearly remains to be done. The first model can be stated as follows:

$$\dot{p} = D(p) - S \tag{6}$$

$$U'(D) = p \tag{7}$$

$$\dot{S} = p - E. \tag{8}$$

The first equation is a very straightforward statement that price p changes at a rate $(\dot{p})$ which depends upon the difference between demand (D) and supply (S). (7) indicates that the level of demand is chosen so that the marginal payoff to resource use (given by $U'(D)$, where $U(D)$ is the payoff function) equals the price. The final equation states that the rate of change of supply depends upon the difference between the present price and the expected future price (E). If the present price is above that expected for the future, supplies are increased, and vice versa. The expectation that the return to holding the commodity will be negative thus encourages holders to move out of it more rapidly, and conversely. This is a mode of behaviour referred to in the discussion above.

The model that we have is clearly a very simple one, and is, in no sense, specific to a resource market, although there is an element of asset-management behaviour underlying (8), and such behaviour is important in resource markets. Nevertheless, the model captures some of the effects referred to earlier, and is also a useful building block. We therefore analyse its solutions. To do this, suppose that $U(D) = \log (D)$. Then (6) becomes

$$\dot{p} = 1/p - S. \tag{6}$$

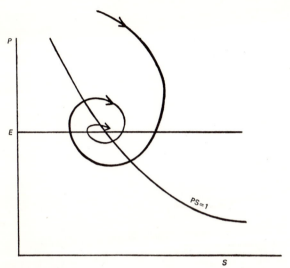

Fig. 2.

The solution to this and (8) depends on whether $E \geqslant 1/\sqrt{2}$ or $E < 1/\sqrt{2}$, and is depicted in Figs. 2 and 3. There is an equilibrium where $1/p = S$ and $p = E$ (demand equals supply and price is at its expected level) and convergence to this will occur, either via damped cycles (as when $E \geqslant 1/\sqrt{2}$) or along a monotone path. The oscillatory behaviour is an illustration of the possibility referred to when we argued that an increase in the current price, if not expected to continue, could lead to a wave of selling and thus raise rather than lower the depletion rate. In this model a high current price generates a belief that the price will fall: there is then a wave of selling that forces the price down to a level which seems lower than an equilibrium value, and then the process starts again.

This model, though suggestive, is clearly far too simple to be convincing, and one of its principal shortcomings is the fact that future price expectations are taken as given. One clearly expects these to be influenced by the history of past prices, and in a market for an exhaustible resource the balance between future supply and demand will also be influential. We therefore augment the earlier system by

$$\dot{E} = a_1(p - E) + a_2 S. \tag{9}$$

The heuristic justification for such an equation is that if price expectations are formed adaptively, one has

$$\dot{E} = a_1(p - E),$$

and if they are a function of cumulative consumption,

$$E = a_2 \int_0^t S(\tau)\,d\tau, \quad \text{or} \quad \dot{E} = a_2 S,$$

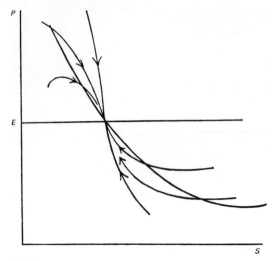

Fig. 3.

and (9) incorporates both of these effects. It is not, of course, the derivative of

$$E = a_1 \varrho \int_0^t p(\tau)\, e^{-(t-\tau)\varrho}\, d\tau + a_2 \int_0^t S(\tau)\, d\tau,$$

as this would involve extra terms in the integrals. We omit these for simplicity. Our model is now:

$$\dot{p} = 1/p - S \tag{6}$$

$$\dot{S} = p - E \tag{8}$$

$$\dot{E} = a_1(p - E) + a_2 S. \tag{9}$$

In spite of the considerable simplifications made in its derivation, this system is not one which is easily analysed. We first consider its equilibria:

$\dot{p} = 0$ i.f.f. $1/p = S$

$\dot{S} = 0$ i.f.f. $p = E$

$\dot{E} = 0$ i.f.f.   either $S = 0$ and $p = E$,   or $-a_1(p - E) = a_2 S$.

Clearly the only stationary point is

$$p = E + \infty, \quad S = 0.$$

This is in fact very reasonable: the only equilibrium is where supply has fallen to zero, and actual and expected prices are equal at infinity. This is certainly what one would expect for a finite resource: endogenising expectations has produced a more believable pattern of long-run behaviour. Unfortunately, it is not easy to make statements about the pattern by which this long-run equilibrium is approached, though it is clear that, as in the earlier case, this may be cyclical. Intuitively one can see this by realising that the present model tends to the earlier one as a_1 and a_2 tend to zero: unless there is a bifurcation in the phase space, one would therefore expect their solution patterns to exhibit some qualitative similarities at least for low values of a_1 and a_2. More formally, one can linearise the system about its stationary point and find the roots of the resulting 3×3 matrix. These are the values of 1 which satisfy the equation

$$1^3 + 1^2 a_1 - 1(1 + a_2) - 2a_1 = 0.$$

This must have one real root, and may have three. It will have only one real root, and hence two complex roots and exhibit cyclical behaviour, if and only if

$$q^2 + (4/27)p^3 > 0 \quad \text{where}$$

$$q = -2a_1 + a_1(a_2 + 1)/3 + 2a_1^3/27 \quad \text{and}$$

$$p = -(1 + a_1) - a_1^3/3. \tag{10}$$

This is a sufficiently complex condition that one cannot easily deduce its implications: however, it is clear that for $a_1 = 0$ (so that $\dot{E} = a_2 S$ and E depends only on cumulative extraction to date), $q = 0$ and $p = -1$, so that there are three real roots and the approach to the solution is monotone, at least locally. The possibility of cyclical behaviour therefore seems to be excluded when expectations are a function only of resource-use and not of past prices. If, on the other hand, $a_2 = 0$, then inequality (10) is satisfied for $a_1 = 1$, but as a_1 is reduced there is a critical value at which the discriminant changes sign. This verifies that if the role of prices relative to quantities in forming expectations is sufficiently great, cyclical behaviour will result. Such a result seems worth having, and very much in keeping with the intuitions developed earlier.

The progression thus far has been from the model (6)–(8) which had little in its structure specifically related to resources, to that of (6a)–(8)–(9) in which expectations were endogenised in a manner which allowed depletion to affect price expectations. We now move to a model which is in some measure an extension of this process: it retains endogenous expectations dependent both on past prices and on cumulative extraction, but models portfolio adjustment in asset markets more directly. It is a natural disequilibrium version of the models of Section IV. We replace the earlier assumption of price adjusting according to the difference between supply and demand by the assumption that prices adjust instantaneously to clear markets, so that

$$D(p) = S \quad \text{or} \quad \frac{\dot{p}}{p} = \frac{-1}{\eta} \frac{\dot{S}}{S}, \tag{11}$$

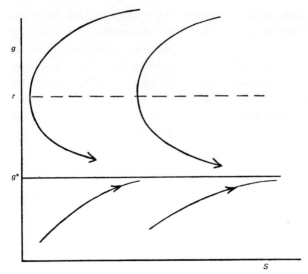

Fig. 4. cη > b.

where η is the elasticity of demand, assumed constant. As before, E will stand for the expected price, and $g = \dot{E}/E$ is the expected rate of capital gain from holding the resource. (8) is now replaced by

$$\frac{\dot{S}}{S} = a(r - g), \tag{12}$$

where r is the rate of return available on other assets. (12) thus implies that the supply of a resource grows rapidly when it is thought to offer a poor return relative to the opportunity cost of capital, and vice versa. The final equation is

$$\frac{\dot{g}}{g} = b\left(\frac{\dot{p}}{p} - g\right) + c\frac{\dot{S}}{S} \tag{13}$$

This has analogies to (9), though now stated in terms of rates of change rather than levels, as befits a model based on ideas of portfolio adjustment. The expected return to holding the resource depends adaptively on past rates of return, and also on the rate of growth of consumption. This model has a clear resemblance to models studied econometrically by Barrow and myself,[1] though it is more amenable than they are to analytical solution.

---

[1] G. M. Heal and M. Barrow, "The influence of interest rates on metal price movements", in *Review of Economic Studies*, 1980 and also G. M. Heal and M. Barrow, in *Economic Letters*, forthcoming and in *Modelling Natural Resource Price Movements: Current Developments in Theory and Practice* (ed. V. K. Smith), forthcoming.

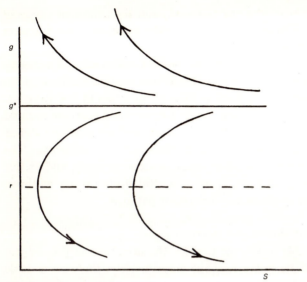

Fig. 5. $c\eta < b$, $ca - (ba/\eta) + b < 0$.

Substituting (11) and (12) into (13) gives

$$\frac{\dot{g}}{g} = (r - g)\left(ca - \frac{ba}{\eta}\right) + bg \tag{14}$$

$$\frac{\dot{S}}{S} = a(r - g). \tag{15}$$

Thus $\dot{g} = 0$, if and only if

$$g = r \bigg/ \left(1 + \frac{b}{ca - ba/\eta}\right) = g^*, \quad \text{say.}$$

Clearly, $g^* > r$ if $c\eta < b$, and vice versa; it is worth noting that if $b = 0$, so that price expectations are a function purely of quantity signals, $g^* = r$ and in equilibrium the expected return to the resource equals that on other assets. In general, the difference between g^* and r tends to zero as b or η shrink to zero. Phase diagrams for the two possible cases are shown in Figs. 4 and 5.

There are actually two possible cases corresponding to $c\eta < b$, the second being when $ca - ba/\eta + b > 0$. However, this case seems uninteresting as $g^* < 0$. Of the two cases shown, the only one with any capacity to mimic an efficient, competitive outcome is Fig. 5 and, in particular, the horizontal path with $g = g^*$ and supply falling at a constant rate $a(r - g^*)$. It is clear that with an isoelastic demand function, as implied by (11), supply will fall at a constant rate in a competitive situation. It is also evident that along the path mentioned, price appreciation is constant at a rate $(-a/\eta)(r - g^*)$, which is, of course, in general neither equal to r nor to g^*, as would be required for efficiency. $\dot{p}/p$

can be shown to be less than g^* along this path, and the persistent excess of expected over actual price increases is caused by the presence of the $\dot{S}/S$ term in the expectations equation.

We thus see that, like its predecessors, the present model will not typically produce an efficient outcome, and that its behaviour is again sensitive to the parameters of the expectations equation. Perhaps one should not have expected anything else: certainly, it seems clear that we could continue to build an almost unlimited range of disequilibrium models, without reaching a clear conclusion about the likely behaviour of markets out of equilibrium. The answer will always depend on the behavioural assumption made, and at this stage it is probably most productive to use empirical studies to try to throw some light on the realism of alternative behaviour patterns.

VI.2. *Conclusion on Expectations and Disequilibrium*

It would be most agreeable to be able to draw the literature on this subject together into an elegant generalisation, but unfortunately, there seems little prospect of being able to do this. In disequilibrium situations with endogenous expectations, a very wide range of outcomes is possible: everything depends on the detailed specification of the model. One is therefore forced to emphasise the need for further empirical work on two issues:

(1) are resource markets in fact significantly out of equilibrium? and
(2) if so, what models of expectations and adjustment are most appropriate?

Research work by Barrow and myself certainly suggests that rates of return on resources and on other assets are not equal, either instantaneously or in the long run. However, there do appear to be forces at work which relate them, though in a way more complex than that predicted by equilibrium theory. However, in general terms work on these topics is just beginning and there is little definitive as yet. Although the theory leaves most issues open, there is one generalisation that seems to have emerged and which is worth emphasising. This is that the stability or otherwise of a resource market does seem to depend on whether expectations about future prices are influenced most by past prices, or by quantity signals such as rates of consumption and remaining stocks. The latter is, perhaps not surprisingly, the more stable case. Remaining stocks can only decrease, and thus concentration on these will tend to impart a steady upward trend to price expectations, as required for efficiency.

VII. Overall Conclusions

I have argued that it is not sensible to ask such questions as: "Are resources scarce?" or "Will the supply of resources meet demand?". The most general question that it seems sensible to ask is: "Are resources being used efficiently?". This is, in intellectual terms, a non-trivial question, and one whose answers carry policy implications. To answer this question it is necessary, but unfortunately not sufficient, to identify how markets would behave if they were to operate efficiently, and to identify sources of inefficiency. This has all been done: what remains is to put numbers into the theory, in order to quantify the efficient path, and then to compare the results with the actual outcomes. Although this is an active research topic at the moment, it seems likely that data problems will ensure that any answer that emerges is, at best, tentative.

If it seems unlikely that we shall be able to decide definitively whether markets have or have not worked efficiently, does this imply that theories of the type expounded in the previous sections are likely to be of little value? Certainly not. One needs a thorough grasp of how efficient markets would work, and of the nature of possible departures from efficiency, in order to evaluate policy proposals. Take as a simple example the proposal to stabilize resource prices, a proposal that has been made in many forms by many agencies in recent years. The theory of the earlier sections is of great value in analysing such proposals. It suggests immediately that we should look very carefully at what is meant by stabilization, and indeed implies that if, as is usually the case, this is identified with holding the price (possibly the real price) constant, then such a policy is likely at best to be unsuccessful and at worst harmful. The point is simply that, as I have shown, equilibrium in resource markets will be characterised by prices that change over time, typically rising. An agency could therefore only hold the price constant if it was willing to buy or sell unlimited amounts. And in the very improbable event that the agency managed to exert a significant influence over prices, then it is doubtful whether this would, in fact, be beneficial. If the pre-intervention equilibrium price trajectory were in fact efficient, then the intervention would more or less by definition be harmful. Obviously, the only intelligent interpretation of the stabilization aim would in this context be stabilization around the efficient equilibrium path, and a prerequisite for the success of such a policy would be a very thorough analytical understanding of the workings of the relevant market.

In the alternative case, when the market before intervention was not working efficiently, it is difficult to see what rationale there could be for stabilization policy or any other systematic intervention. Of course, the natural response is that it should be aimed to restore efficiency, but whether this lies within the scope of market management policies will depend very much on the cause of the inefficiency. One can hardly expect inefficiencies caused by monopoly or imperfect competition to be removed in this way, and even those caused by uncertainties and imperfect information, while they may be ameliorated by careful intervention designed to reduce fluctuations, are surely best tackled by research and forecasting designed to reduce the uncertainties.